Ben Stacy Jerrik (Ed.)

Sequence of Events Recorder

Ben Stacy Jerrik (Ed.)

Sequence of Events Recorder

Global Positioning System, Root cause analysis

Part Press

Imprint

Publisher:
Part Press is a trademark of
International Book Market Service Ltd., 17 Rue Meldrum, Beau Bassin, 1713-01 Mauritius
Email: info@bookmarketservice.com
Website: www.bookmarketservice.com

Published in 2013

Printed in: U.S.A., U.K., Germany. This book was not produced in Mauritius.

ISBN: 978-613-9-11865-6

Contents

Articles

References

Sequence_of_events_recorder

A **sequence of events recorder** (SER)[1] is an intelligent standalone microprocessor based system, which monitors external inputs and records the time and sequence of the changes.[2] [3] Sequence of events recorders usually have an external time source such as a GPS or radio clock. When wired inputs change state the time and state of each change is recorded.

SERs enable rapid root cause analysis after multiple events have occurred due to the secure recording of the sequence of events in the order of occurrence. SERs are therefore utilized as a diagnostic tool to minimize plant downtime. SERs are often interfaced with a SCADA system, distributed control system (DCS),or programmable logic controller (PLC).

SER reports are used by electrical engineers to analyze large and small electrical system blackouts. After the Northeast blackout of 2003, the North American Electric Reliability Corporation (NERC) specified that electrical system data should be time-tagged to the nearest millisecond.

In 1984 The Tetragenics Company, a subsidiary of the Montana Power Company, introduced the first remote terminal unit (RTU) that time-tagged events to the nearest millisecond, and now there are also other RTUs with this capability. Digital protective relays and some PLCs now also include time-tagging to the nearest millisecond; SCADA systems that incorporate these devices provide SER functions without a dedicated SER device.

See also

- Data logger

References

[1] "Sequence of events recorder" (http://acronyms.thefreedictionary.com/Sequence+of+events+recorder). The Free Dictionary by Farlex. .

[2] "Sequence of events recorder facility for an industrial process control ..." (http://www.google.com/patents/about?id=vM_bAAAAEBAJ). Google Patents. 2010. .

[3] "Utility Communications Architecture (UCA) glossary" (http://www.nettedautomation.com/glossary_menue/glossary_uca.html). NettedAutomation. .

Global_Positioning_System

The **Global Positioning System** (**GPS**) is a space-based satellite navigation system that provides location and time information in all weather conditions, anywhere on or near the Earth where there is an unobstructed line of sight to four or more GPS satellites. The system provides critical capabilities to military, civil and commercial users around the world. It is maintained by the United States government and is freely accessible to anyone with a GPS receiver.

Artist's conception of GPS Block II-F satellite in Earth orbit.

The GPS project was developed in 1973 to overcome the limitations of previous navigation systems,[1] integrating ideas from several predecessors, including a number of classified engineering design studies from the 1960s. GPS was created and realized by the U.S. Department of Defense (DoD) and was originally run with 24 satellites. It became fully operational in 1994.

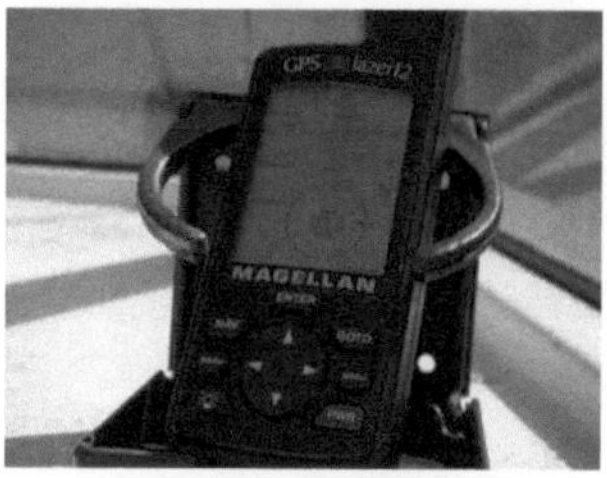

Civilian GPS receiver ("GPS navigation device") in a marine application.

Advances in technology and new demands on the existing system have now led to efforts to modernize the GPS system and implement the next generation of GPS III satellites and Next Generation Operational Control System (OCX).[2] Announcements from the Vice President and the White House in 1998 initiated these changes. In 2000, U.S. Congress authorized the modernization effort, referred to as GPS III.

In addition to GPS, other systems are in use or under development. The Russian Global Navigation Satellite System (GLONASS) was developed contemporaneously with GPS, but suffered from incomplete coverage of the globe until the mid-2000s. There are also the planned European Union Galileo positioning system, Chinese Compass navigation system, and Indian Regional Navigational Satellite System.

Automotive navigation system in a taxicab.

History

The design of GPS is based partly on similar ground-based radio-navigation systems, such as LORAN and the Decca Navigator developed in the early 1940s, and used during World War II.

Predecessors

In 1956, the German-American physicist Friedwardt Winterberg[3] proposed a test of general relativity (for time slowing in a strong gravitational field) using accurate atomic clocks placed in orbit inside artificial satellites. (To achieve accuracy requirements, GPS uses principles of general relativity to correct the satellites' atomic clocks.[4]) Additional inspiration for GPS came when the Soviet Union launched the first man-made satellite, Sputnik in 1957. Two American physicists, William Guier and George Weiffenbach, at Johns Hopkins's

Applied Physics Laboratory (APL), decided on their own to monitor Sputnik's radio transmissions.[5] Within hours they realized that, because of the Doppler effect, they could pinpoint where the satellite was along its orbit from the Doppler shift. The Director of the APL gave them access to their UNIVAC to do the heavy calculations required. The following spring, Frank McClure, the deputy director of the APL, asked Guier and Weiffenbach to investigate the inverse problem - pinpointing the user's location given that of the satellite. (The Navy was developing the submarine-launched Polaris missile, which required them to know the submarine's location.) This led them and APL to develop the Transit system.[6]

GPS receivers are now integrated in many mobile phones.

U.S. Air Force Senior Airman runs through a checklist during Global Positioning System satellite operations.

Official logo for NAVSTAR GPS Emblem of the 50th Space Wing

The first satellite navigation system, Transit, used by the United States Navy, was first successfully tested in 1960. It used a constellation of five satellites and could provide a navigational fix approximately once per hour. In 1967, the U.S. Navy developed the Timation satellite that proved the ability to place accurate clocks in space, a technology required by GPS. In the 1970s, the ground-based Omega Navigation System, based on phase comparison of signal transmission from pairs of stations,[7] became the first worldwide radio navigation system. Limitations of these systems drove the need for a more universal navigation solution with greater accuracy.

While there were wide needs for accurate navigation in military and civilian sectors, almost none of those were seen as justification for the billions of dollars it would cost in research, development, deployment, and operation for a constellation of navigation satellites. During the Cold War arms race, the nuclear threat to the existence of the United States was the one need that did justify this cost in the view of the United States Congress. This deterrent effect is why GPS was funded. It is also the reason for the ultra secrecy at that time. The nuclear triad consisted of the United States Navy's submarine-launched ballistic missiles (SLBMs) along with United States Air Force (USAF) strategic bombers and intercontinental ballistic missiles (ICBMs). Considered vital to the nuclear deterrence posture, accurate determination of the SLBM launch position was a force multiplier.

Precise navigation would enable United States submarines to get an accurate fix of their positions prior to launching their SLBMs.[8] The USAF with two-thirds of the nuclear triad also had requirements for a more accurate and reliable navigation system. The Navy and Air Force were developing their own technologies in parallel to solve what was essentially the same problem. To increase the survivability of ICBMs, there was a proposal to use mobile launch platforms (such as Russian SS-24 and SS-25) and so the need to fix the launch position had similarity to the SLBM situation.

In 1960, the Air Force proposed a radio-navigation system called MOSAIC (MObile System for Accurate ICBM Control) that was essentially a 3-D LORAN. A follow-on study called Project 57 was worked in 1963 and it was "in this study that the GPS concept was born." That same year the concept was pursued as Project 621B, which had "many of the attributes that you now see in GPS"[9] and promised increased accuracy for Air Force bombers as well as ICBMs. Updates from the Navy Transit system were too slow for the high speeds of Air Force operation. The Navy Research Laboratory continued advancements with their Timation (Time Navigation) satellites, first launched in 1967, and with the third one in 1974 carrying the first atomic clock into orbit.[10]

Another important predecessor to GPS came from a different branch of the United States military. In 1964, the United States Army orbited its first SECOR (Sequential Collation of Range) satellite used for geodetic surveying. The SECOR system included three ground-based transmitters from known locations that would send signals to the satellite transponder in orbit. A fourth ground-based station at an undetermined position could then use those signals to fix its location precisely. The last SECOR satellite was launched in 1969.[11] Decades later during the early years of GPS, civilian surveying became one of the first fields to make use of the new technology, because they could reap benefits of signals from the less-than-complete GPS constellation years before it was declared operational. GPS can be thought of as an evolution of the SECOR system where the ground-based transmitters have been migrated into orbit.

Development

With these parallel developments in the 1960s, it was realized that a superior system could be developed by synthesizing the best technologies from 621B, Transit, Timation, and SECOR in a multi-service program.

During Labor Day weekend in 1973, a meeting of about 12 military officers at the Pentagon discussed the creation of a *Defense Navigation Satellite System (DNSS)*. It was at this meeting that "the real synthesis that became GPS was created." Later that year, the DNSS program was named *Navstar*. With the individual satellites being associated with the name Navstar (as with the predecessors Transit and Timation), a more fully encompassing name was used to identify the constellation of Navstar satellites, *Navstar-GPS*, which was later shortened simply to GPS.[12]

After Korean Air Lines Flight 007, carrying 269 people, was shot down in 1983 after straying into the USSR's prohibited airspace,[13] in the vicinity of Sakhalin and Moneron Islands, President Ronald Reagan issued a directive making GPS freely available for civilian use, once it was sufficiently developed, as a common good.[14] The first satellite was launched in 1989, and the 24th satellite was launched in 1994.

Initially, the highest quality signal was reserved for military use, and the signal available for civilian use was intentionally degraded (Selective Availability). This changed with President Bill Clinton ordering Selective Availability to be turned off at midnight May 1, 2000, improving the precision of civilian GPS from 100 meters (ft) to 20 meters (ft). The executive order signed in 1996 to turn off Selective Availability in 2000 was proposed by the US Secretary of Defense, William Perry, because of the widespread growth of differential GPS services to improve civilian accuracy and eliminate the US military advantage. Moreover, the US military was actively developing technologies to deny GPS service to potential adversaries on a regional basis.[15]

Over the last decade, the U.S. has implemented several improvements to the GPS service, including new signals for civil use and increased accuracy and integrity for all users, all while maintaining compatibility with existing GPS equipment.

GPS modernization[16] has now become an ongoing initiative to upgrade the Global Positioning System with new capabilities to meet growing military, civil, and commercial needs. The program is being implemented through a series of satellite acquisitions, including GPS Block III and the Next Generation Operational Control System (OCX).

The U.S. Government continues to improve the GPS space and ground segments to increase performance and accuracy.

GPS is owned and operated by the United States Government as a national resource. Department of Defense (DoD) is the steward of GPS. *Interagency GPS Executive Board (IGEB)* oversaw GPS policy matters from 1996 to 2004. After that the National Space-Based Positioning, Navigation and Timing Executive Committee was established by presidential directive in 2004 to advise and coordinate federal departments and agencies on matters concerning the GPS and related systems. The executive committee is chaired jointly by the deputy secretaries of defense and transportation. Its membership includes equivalent-level officials from the departments of state, commerce, and homeland security, the joint chiefs of staff, and NASA. Components of the executive office of the president participate as observers to the executive committee, and the FCC chairman participates as a liaison.

The DoD is required by law to "maintain a Standard Positioning Service (as defined in the federal radio navigation plan and the standard positioning service signal specification) that will be available on a continuous, worldwide basis," and "develop measures to prevent hostile use of GPS and its augmentations without unduly disrupting or degrading civilian uses."

Timeline and modernization

Summary of satellites[17]

Block	Launch Period	Satellite launches				Currently in orbit and healthy
		Suc-cess	Fail-ure	In prep-aration	Plan-ned	
I	1978–1985	10	1	0	0	0
II	1989–1990	9	0	0	0	0
IIA	1990–1997	19	0	0	0	9
IIR	1997–2004	12	1	0	0	12
IIR-M	2005–2009	8	0	0	0	7
IIF	From 2010	3	0	10	0	2
IIIA	From 2014	0	0	0	12	0
IIIB	—	0	0	0	8	0
IIIC	—	0	0	0	16	0
Total		61	2	10	36	30

(Last update: October 8, 2012)
PRN 01 from Block IIR-M is unhealthy
PRN 25 from Block IIA is unhealthy
PRN 32 from Block IIA is unhealthy
PRN 27 from Block IIA is unhealthy
[18] For a more complete list, see *list of GPS satellite launches*

- In 1972, the USAF Central Inertial Guidance Test Facility (Holloman AFB), conducted developmental flight tests of two prototype GPS receivers over White Sands Missile Range, using ground-based pseudo-satellites.
- In 1978, the first experimental Block-I GPS satellite was launched.
- In 1983, after Soviet interceptor aircraft shot down the civilian airliner KAL 007 that strayed into prohibited airspace because of navigational errors, killing all 269 people on board, U.S. President Ronald Reagan announced that GPS would be made available for civilian uses once it was completed,[19] [20] although it had been previously

published [in Navigation magazine] that the CA code (Coarse Acquisition code) would be available to civilian users.

- By 1985, ten more experimental Block-I satellites had been launched to validate the concept. Command & Control of these satellites had moved from Onizuka AFS, CA and turned over to the 2nd Satellite Control Squadron (2SCS) located at Falcon Air Force Station in Colorado Springs, Colorado.[21] [22]
- On February 14, 1989, the first modern Block-II satellite was launched.
- The Gulf War from 1990 to 1991, was the first conflict where GPS was widely used.[23]
- In 1992, the 2nd Space Wing, which originally managed the system, was de-activated and replaced by the 50th Space Wing.
- By December 1993, GPS achieved initial operational capability (IOC), indicating a full constellation (24 satellites) was available and providing the Standard Positioning Service (SPS).[24]
- Full Operational Capability (FOC) was declared by Air Force Space Command (AFSPC) in April 1995, signifying full availability of the military's secure Precise Positioning Service (PPS).[24]
- In 1996, recognizing the importance of GPS to civilian users as well as military users, U.S. President Bill Clinton issued a policy directive[25] declaring GPS to be a dual-use system and establishing an Interagency GPS Executive Board to manage it as a national asset.
- In 1998, United States Vice President Al Gore announced plans to upgrade GPS with two new civilian signals for enhanced user accuracy and reliability, particularly with respect to aviation safety and in 2000 the United States Congress authorized the effort, referring to it as *GPS III*.
- On May 2, 2000 "Selective Availability" was discontinued as a result of the 1996 executive order, allowing users to receive a non-degraded signal globally.
- In 2004, the United States Government signed an agreement with the European Community establishing cooperation related to GPS and Europe's planned Galileo system.
- In 2004, United States President George W. Bush updated the national policy and replaced the executive board with the National Executive Committee for Space-Based Positioning, Navigation, and Timing.[26]
- November 2004, Qualcomm announced successful tests of assisted GPS for mobile phones.[27]
- In 2005, the first modernized GPS satellite was launched and began transmitting a second civilian signal (L2C) for enhanced user performance.
- On September 14, 2007, the aging mainframe-based Ground Segment Control System was transferred to the new Architecture Evolution Plan.[28]
- On May 19, 2009, the United States Government Accountability Office issued a report warning that some GPS satellites could fail as soon as 2010.[29]
- On May 21, 2009, the Air Force Space Command allayed fears of GPS failure saying "There's only a small risk we will not continue to exceed our performance standard."[30]
- On January 11, 2010, an update of ground control systems caused a software incompatibility with 8000 to 10000 military receivers manufactured by a division of Trimble Navigation Limited of Sunnyvale, Calif.[31]
- On February 25, 2010,[32] the U.S. Air Force awarded the contract to develop the GPS Next Generation Operational Control System (OCX) to improve accuracy and availability of GPS navigation signals, and serve as a critical part of GPS modernization.
- A GPS satellite was launched on May 28, 2010.[33] The oldest GPS satellite still in operation was launched on November 26, 1990, and became operational on December 10, 1990.[34]
- The GPS satellite, GPS IIF-2, was launched on July 16, 2011 at 06:41 GMT from Space Launch Complex 37B at the Cape Canaveral Air Force Station.[35]
- The GPS satellite, GPS IIF-3, was launched on October 4, 2012 at 12:10 GMT from Space Launch Complex 37B at the Cape Canaveral Air Force Station.[36]

Awards

On February 10, 1993, the National Aeronautic Association selected the GPS Team as winners of the 1992 Robert J. Collier Trophy, the nation's most prestigious aviation award. This team combines researchers from the Naval Research Laboratory, the USAF, the Aerospace Corporation, Rockwell International Corporation, and IBM Federal Systems Company. The citation honors them "for the most significant development for safe and efficient navigation and surveillance of air and spacecraft since the introduction of radio navigation 50 years ago."

Two GPS developers received the National Academy of Engineering Charles Stark Draper Prize for 2003:

- Ivan Getting, emeritus president of The Aerospace Corporation and an engineer at the Massachusetts Institute of Technology, established the basis for GPS, improving on the World War II land-based radio system called LORAN (*Lo*ng-range *R*adio *A*id to *N*avigation).
- Bradford Parkinson, professor of aeronautics and astronautics at Stanford University, conceived the present satellite-based system in the early 1960s and developed it in conjunction with the U.S. Air Force. Parkinson served twenty-one years in the Air Force, from 1957 to 1978, and retired with the rank of colonel.
- GPS developer Roger L. Easton received the National Medal of Technology on February 13, 2006.[37]

In 1998, GPS technology was inducted into the Space Foundation Space Technology Hall of Fame.[38]

Francis X. Kane (Col. USAF, ret.) was inducted into the U.S. Air Force Space and Missile Pioneers Hall of Fame at Lackland A.F.B., San Antonio, Texas, March 2, 2010 for his role in space technology development and the engineering design concept of GPS conducted as part of Project 621B.

On October 4, 2011, the International Astronautical Federation (IAF) awarded the Global Positioning System (GPS) its 60th Anniversary Award, nominated by IAF member, the American Institute for Aeronautics and Astronautics (AIAA). The IAF Honors and Awards Committee recognized the uniqueness of the GPS program and the exemplary role it has played in building international collaboration for the benefit of humanity.

Basic concept of GPS

A GPS receiver calculates its position by precisely timing the signals sent by GPS satellites high above the Earth. Each satellite continually transmits messages that include

- the time the message was transmitted
- satellite position at time of message transmission

The receiver uses the messages it receives to determine the transit time of each message and computes the distance to each satellite using the speed of light. Each of these distances and satellites' locations define a sphere. The receiver is on the surface of each of these spheres when the distances and the satellites' locations are correct. These distances and satellites' locations are used to compute the location of the receiver using the navigation equations. This location is then displayed, perhaps with a moving map display or latitude and longitude; elevation information may be included. Many GPS units show derived information such as direction and speed, calculated from position changes.

In typical GPS operation, four or more satellites must be visible to obtain an accurate result. Four sphere surfaces typically do not intersect. [39] Because of this we can say with confidence that when we solve the navigation equations to find an intersection, this solution gives us the position of the receiver along with accurate time thereby eliminating the need for a very large, expensive, and power hungry clock. The very accurately computed time is used only for display or not at all in many GPS applications, which use only the location. A number of applications for GPS do make use of this cheap and highly accurate timing. These include time transfer, traffic signal timing, and synchronization of cell phone base stations.

Although four satellites are required for normal operation, fewer apply in special cases. If one variable is already known, a receiver can determine its position using only three satellites. For example, a ship or aircraft may have known elevation. Some GPS receivers may use additional clues or assumptions such as reusing the last known altitude, dead reckoning, inertial navigation, or including information from the vehicle computer, to give a (possibly degraded) position when fewer than four satellites are visible.[40] [41] [42]

Structure

The current GPS consists of three major segments. These are the space segment (SS), a control segment (CS), and a user segment (US).[43] The U.S. Air Force develops, maintains, and operates the space and control segments. GPS satellites broadcast signals from space, and each GPS receiver uses these signals to calculate its three-dimensional location (latitude, longitude, and altitude) and the current time.[44]

The space segment is composed of 24 to 32 satellites in medium Earth orbit and also includes the payload adapters to the boosters required to launch them into orbit. The control segment is composed of a master control station, an alternate master control station, and a host of dedicated and shared ground antennas and monitor stations. The user segment is composed of hundreds of thousands of U.S. and allied military users of the secure GPS Precise Positioning Service, and tens of millions of civil, commercial, and scientific users of the Standard Positioning Service (see GPS navigation devices).

Space segment

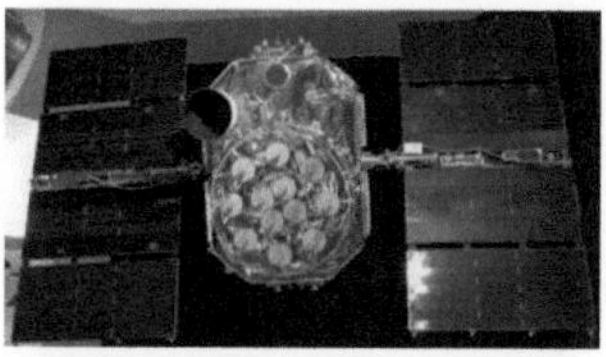

Unlaunched GPS block II-A satellite on display at the San Diego Air & Space Museum

The space segment (SS) is composed of the orbiting GPS satellites, or Space Vehicles (SV) in GPS parlance. The GPS design originally called for 24 SVs, eight each in three approximately circular orbits,[45] but this was modified to six orbital planes with four satellites each.[46] The orbits are centered on the Earth, not rotating with the Earth, but instead fixed with respect to the distant stars.[47] The six orbit planes have approximately 55° inclination (tilt relative to Earth's equator) and are separated by 60° right ascension of the ascending node (angle along the equator from a reference point to the orbit's intersection).[48] The orbital period is one-half a sidereal day, i.e., 11 hours and 58 minutes.[49] The orbits are arranged so that at least six satellites are always within line of sight from almost everywhere on Earth's surface.[50] The result of this objective is that the four satellites are not evenly spaced (90 degrees) apart within each orbit. In general terms, the angular difference between satellites in each orbit is 30, 105, 120, and 105 degrees apart which, of course, sum to 360 degrees.

Orbiting at an altitude of approximately 20200 km (mi); orbital radius of approximately 26600 km (mi), each SV makes two complete orbits each sidereal day, repeating the same ground track each day.[51] This was very helpful during development because even with only four satellites, correct alignment means all four are visible from one spot for a few hours each day. For military operations, the ground track repeat can be used to ensure good coverage in combat zones.

As of December 2012,[52] there are 32 satellites in the GPS constellation. The additional satellites improve the precision of GPS receiver calculations by providing redundant measurements. With the increased number of satellites, the constellation was changed to a nonuniform arrangement. Such an arrangement was shown to improve reliability and availability of the system, relative to a uniform system, when multiple satellites fail.[53] About nine satellites are visible from any point on the ground at any one time (see animation at right), ensuring considerable redundancy over the minimum four satellites needed for a position.

Control segment

The control segment is composed of

1. a master control station (MCS),
2. an alternate master control station,
3. four dedicated ground antennas and
4. six dedicated monitor stations

Ground monitor station used from 1984 to 2007, on display at the Air Force Space & Missile Museum

The MCS can also access U.S. Air Force Satellite Control Network (AFSCN) ground antennas (for additional command and control capability) and NGA (National Geospatial-Intelligence Agency) monitor stations. The flight paths of the satellites are tracked by dedicated U.S. Air Force monitoring stations in Hawaii, Kwajalein, Ascension Island, Diego Garcia, Colorado Springs, Colorado and Cape Canaveral, along with shared NGA monitor stations operated in England, Argentina, Ecuador, Bahrain, Australia and Washington DC.[54] The tracking information is sent to the Air Force Space Command MCS at Schriever Air Force Base 25 km (mi) ESE of Colorado Springs, which is operated by the 2nd Space Operations Squadron (2 SOPS) of the U.S. Air Force. Then 2 SOPS contacts each GPS satellite regularly with a navigational update using dedicated or shared (AFSCN) ground antennas (GPS dedicated ground antennas are located at Kwajalein, Ascension Island, Diego Garcia, and Cape Canaveral). These updates synchronize the atomic clocks on board the satellites to within a few nanoseconds of each other, and adjust the ephemeris of each satellite's internal orbital model. The updates are created by a Kalman filter that uses inputs from the ground monitoring stations, space weather information, and various other inputs.[55]

Satellite maneuvers are not precise by GPS standards. So to change the orbit of a satellite, the satellite must be marked *unhealthy*, so receivers will not use it in their calculation. Then the maneuver can be carried out, and the resulting orbit tracked from the ground. Then the new ephemeris is uploaded and the satellite marked healthy again.

The Operation Control Segment (OCS) currently serves as the control segment of record. It provides the operational capability that supports global GPS users and keeps the GPS system operational and performing within specification.

OCS successfully replaced the legacy 1970's-era mainframe computer at Schriever Air Force Base in September 2007. After installation, the system helped enable upgrades and provide a foundation for a new security architecture that supported the U.S. armed forces. OCS will continue to be the ground control system of record until the new segment, Next Generation GPS Operation Control System[2] (OCX), is fully developed and functional.

The new capabilities provided by OCX will be the cornerstone for revolutionizing GPS's mission capabilities, and enabling[56] Air Force Space Command to greatly enhance GPS operational services to U.S. combat forces, civil partners and myriad of domestic and international users.

The GPS OCX program also will reduce cost, schedule and technical risk. It is designed to provide 50%[57] sustainment cost savings through efficient software architecture and Performance-Based Logistics. In addition, GPS OCX expected to cost millions less than the cost to upgrade OCS while providing four times the capability.

The GPS OCX program represents a critical part of GPS modernization and provides significant information assurance improvements over the current GPS OCS program.

- OCX will have the ability to control and manage GPS legacy satellites as well as the next generation of GPS III satellites, while enabling the full array of military signals.
- Built on a flexible architecture that can rapidly adapt to the changing needs of today's and future GPS users allowing immediate access to GPS data and constellations status through secure, accurate and reliable information.
- Empowers the warfighter with more secure, actionable and predictive information to enhance situational awareness.

- Enables new modernized signals (L1C, L2C, and L5) and has M-code capability, which the legacy system is unable to do.
- Provides significant information assurance improvements over the current program including detecting and preventing cyber attacks, while isolating, containing and operating during such attacks.
- Supports higher volume near real-time command and control capabilities.

On September 14, 2011,[58] the U.S. Air Force announced the completion of GPS OCX Preliminary Design Review and confirmed that the OCX program is ready for the next phase of development.

The GPS OCX program has achieved major milestones and is on track to support the GPS IIIA launch in May 2014.

User segment

The user segment is composed of hundreds of thousands of U.S. and allied military users of the secure GPS Precise Positioning Service, and tens of millions of civil, commercial and scientific users of the Standard Positioning Service. In general, GPS receivers are composed of an antenna, tuned to the frequencies transmitted by the satellites, receiver-processors, and a highly stable clock (often a crystal oscillator). They may also include a display for providing location and speed information to the user. A receiver is often described by its number of channels: this signifies how many satellites it can monitor simultaneously. Originally limited to four or five, this has progressively increased over the years so that, as of 2007, receivers typically have between 12 and 20 channels.[59]

GPS receivers come in a variety of formats, from devices integrated into cars, phones, and watches, to dedicated devices such as these.

GPS receivers may include an input for differential corrections, using the RTCM SC-104 format. This is typically in the form of an RS-232 port at 4,800 bit/s speed. Data is actually sent at a much lower rate, which limits the accuracy of the signal sent using RTCM . Receivers with internal DGPS receivers can outperform those using external RTCM data . As of 2006, even low-cost units commonly include Wide Area Augmentation System (WAAS) receivers.

A typical OEM GPS receiver module measuring 15×17 mm.

Many GPS receivers can relay position data to a PC or other device using the NMEA 0183 protocol. Although this protocol is officially defined by the National Marine Electronics Association (NMEA),[60] references to this protocol have been compiled from public records, allowing open source tools like gpsd to read the protocol without violating intellectual property laws. Other proprietary protocols exist as well, such as the SiRF and MTK protocols. Receivers can interface with other devices using methods including a serial connection, USB, or Bluetooth.

Applications

While originally a military project, GPS is considered a *dual-use* technology, meaning it has significant military and civilian applications.

GPS has become a widely deployed and useful tool for commerce, scientific uses, tracking, and surveillance. GPS's accurate time facilitates everyday activities such as banking, mobile phone operations, and even the control of power grids by allowing well synchronized hand-off switching.[44]

Civilian

Many civilian applications use one or more of GPS's three basic components: absolute location, relative movement, and time transfer.

This antenna is mounted on the roof of a hut containing a scientific experiment needing precise timing.

- Cartography: Both civilian and military cartographers use GPS extensively.
- Cellular telephony: Clock synchronization enables time transfer, which is critical for synchronizing its spreading codes with other base stations to facilitate inter-cell handoff and support hybrid GPS/cellular position detection for mobile emergency calls and other applications. The first handsets with integrated GPS launched in the late 1990s. The U.S. Federal Communications Commission (FCC) mandated the feature in either the handset or in the towers (for use in triangulation) in 2002 so emergency services could locate 911 callers. Third-party software developers later gained access to GPS APIs from Nextel upon launch, followed by Sprint in 2006, and Verizon soon thereafter.
- Clock synchronization: The accuracy of GPS time signals (±10 ns)[61] is second only to the atomic clocks upon which they are based.
- Disaster relief/emergency services: Depend upon GPS for location and timing capabilities.
- Fleet Tracking: The use of GPS technology to identify, locate and maintain contact reports with one or more fleet vehicles in real-time.
- Geofencing: Vehicle tracking systems, person tracking systems, and pet tracking systems use GPS to locate a vehicle, person, or pet. These devices are attached to the vehicle, person, or the pet collar. The application provides continuous tracking and mobile or Internet updates should the target leave a designated area.[62]
- Geotagging: Applying location coordinates to digital objects such as photographs and other documents for purposes such as creating map overlays.
- GPS Aircraft Tracking
- GPS tours: Location determines what content to display; for instance, information about an approaching point of interest.
- Navigation: Navigators value digitally precise velocity and orientation measurements.
- Phasor measurements: GPS enables highly accurate timestamping of power system measurements, making it possible to compute phasors.
- Recreation: For example, geocaching, geodashing, GPS drawing and waymarking.
- Robotics: Self-navigating, autonomous robots using a GPS sensors, which calculate latitude, longitude, time, speed, and heading.
- Surveying: Surveyors use absolute locations to make maps and determine property boundaries.
- Tectonics: GPS enables direct fault motion measurement in earthquakes.
- Telematics: GPS technology integrated with computers and mobile communications technology in automotive navigation systems

Restrictions on civilian use

The U.S. Government controls the export of some civilian receivers. All GPS receivers capable of functioning above 18 kilometres (mi) altitude and 515 metres per second (kn) or designed, modified for use with unmanned air vehicles like e.g. ballistic or cruise missile systems are classified as munitions (weapons) for which State Department export licenses are required.[63]

This rule applies even to otherwise purely civilian units that only receive the L1 frequency and the C/A (Coarse/Acquisition) code and cannot correct for Selective Availability (SA), etc.

Disabling operation above these limits exempts the receiver from classification as a munition. Vendor interpretations differ. The rule refers to operation at both the target altitude and speed, but some receivers stop operating even when stationary. This has caused problems with some amateur radio balloon launches that regularly reach 30 kilometres (mi).

These limits only apply to units exported from (or which have components exported from) the USA - there is a growing trade in various components, including GPS units, supplied by other countries, which are expressly sold as ITAR-free.

Military

As of 2009, military applications of GPS include:

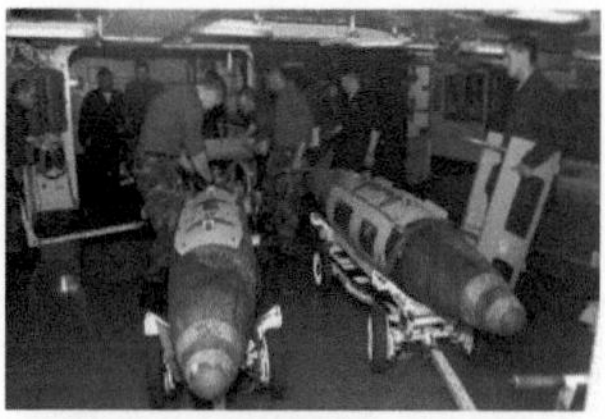

Attaching a GPS guidance kit to a 'dumb' bomb, March 2003.

- Navigation: GPS allows soldiers to find objectives, even in the dark or in unfamiliar territory, and to coordinate troop and supply movement. In the United States armed forces, commanders use the *Commanders Digital Assistant* and lower ranks use the *Soldier Digital Assistant*.[64] [65] [66] [67]
- Target tracking: Various military weapons systems use GPS to track potential ground and air targets before flagging them as hostile. These weapon systems pass target coordinates to precision-guided munitions to allow them to engage targets accurately. Military aircraft, particularly in air-to-ground roles, use GPS to find targets (for example, gun camera video from AH-1 Cobras in Iraq show GPS co-ordinates that can be viewed with specialized software).
- Missile and projectile guidance: GPS allows accurate targeting of various military weapons including ICBMs, cruise missiles, precision-guided munitions and Artillery projectiles. Embedded GPS receivers able to withstand accelerations of 12,000 *g* or about 118 km/s^2 have been developed for use in 155 millimetres (in) howitzers.[68]
- Search and Rescue: Downed pilots can be located faster if their position is known.
- Reconnaissance: Patrol movement can be managed more closely.
- GPS satellites carry a set of nuclear detonation detectors consisting of an optical sensor (Y-sensor), an X-ray sensor, a dosimeter, and an electromagnetic pulse (EMP) sensor (W-sensor), that form a major portion of the United States Nuclear Detonation Detection System.[69] [70]

Communication

The navigational signals transmitted by GPS satellites encode a variety of information including satellite positions, the state of the internal clocks, and the health of the network. These signals are transmitted on two separate carrier frequencies that are common to all satellites in the network. Two different encodings are used: a public encoding that enables lower resolution navigation, and an encrypted encoding used by the U.S. military.

Message format

GPS message format

Subframes	Description
1	Satellite clock, GPS time relationship
2–3	Ephemeris (precise satellite orbit)
4–5	Almanac component (satellite network synopsis, error correction)

Each GPS satellite continuously broadcasts a *navigation message* on L1 C/A and L2 P/Y frequencies at a rate of 50 bits per second (see bitrate). Each complete message takes 750 seconds (12 1/2 minutes) to complete. The message structure has a basic format of a 1500-bit-long frame made up of five subframes, each subframe being 300 bits (6 seconds) long. Subframes 4 and 5 are subcommutated 25 times each, so that a complete data message requires the transmission of 25 full frames. Each subframe consists of ten words, each 30 bits long. Thus, with 300 bits in a subframe times 5 subframes in a frame times 25 frames in a message, each message is 37,500 bits long. At a transmission rate of 50 bps, this gives 750 seconds to transmit an entire almanac message. Each 30-second frame begins precisely on the minute or half-minute as indicated by the atomic clock on each satellite.[71]

The first subframe of each frame encodes the week number and the time within the week,[72] as well as the data about the health of the satellite. The second and the third subframes contain the *ephemeris* - the precise orbit for the satellite. The fourth and fifth subframes contain the *almanac*, which contains coarse orbit and status information for up to 32 satellites in the constellation as well as data related to error correction. Thus, in order to obtain an accurate satellite location from this transmitted message the receiver must demodulate the message from each satellite it includes in its solution for 18 to 30 seconds. In order to collect all the transmitted almanacs the receiver must demodulate the message for 732 to 750 seconds or 12 1/2 minutes.[73]

All satellites broadcast at the same frequencies. Signals are encoded using code division multiple access (CDMA) allowing messages from individual satellites to be distinguished from each other based on unique encodings for each satellite (that the receiver must be aware of). Two distinct types of CDMA encodings are used: the coarse/acquisition (C/A) code, which is accessible by the general public, and the precise P(Y) code, which is encrypted so that only the U.S. military can access it.

The ephemeris is updated every 2 hours and is generally valid for 4 hours, with provisions for updates every 6 hours or longer in non-nominal conditions. The almanac is updated typically every 24 hours. Additionally data for a few weeks following is uploaded in case of transmission updates that delay data upload.

Subframe #	Page #	Name	Word #	Bits	Scale	Signed
1	all	Week Number	3	1-10	1:1	No
1	all	CA or P On L2	3	11,12	1:1	No
1	all	URA Index	3	13-16	1:1	No
1	all	SV_Health	3	17-22	1:1	No
1	all	IODC(MSB)	3	23,24	1:1	No
1	all	L2Pdata flag	4	1	1:1	No
1	all	ResW4	4	2-24	N/A	N/A
1	all	ResW5	5	1-24	N/A	N/A
1	all	ResW6	6	1-24	N/A	N/A
1	all	ResW7	7	1-16	N/A	N/A
1	all	TGD	7	17-24	2^-31	Yes
1	all	IODC (LSB)	8	1-8	1:1	No
1	all	TOC	8	9-24	2^4	No
1	all	AF2	9	1-8	2^-55	Yes
1	all	AF1	9	9-24	2^-43	Yes
1	all	AF0	10	1-22	2^-31	Yes

Subframe #	Page #	Name	Word #	Bits	Scale	Signed
2	all	IODE	3	1-8	1:1	No
2	all	CRS	3	9-24	2^-5	Yes
2	all	Delta N	4	1-16	2^-43	Yes
2	all	M0 (MSB)	4	17-24	2^-31	Yes
2	all	M0 (LSB)	5	1-24		
2	all	CUC	6	1-16	2^-29	Yes
2	all	e (MSB)	6	17-24	2^-33	No
2	all	e (LSB)	7	1-24		
2	all	CUS	8	1-16	2^-29	Yes
2	all	root A (MSB)	8	17-24	2^-19	No
2	all	root A (LSB)	9	1-24		
2	all	TOE	10	1-16	2^4	No
2	all	FitInt	10	17	1:1	No
2	all	AODO	10	18-22	900	No

Subframe #	Page #	Name	Word #	Bits	Scale	Signed
3	all	CIC	3	1-16	2^-29	Yes
3	all	Omega 0 (MSB)	3	17-24	2^-31	Yes
3	all	Omega 0 (LSB)	4	1-24		
3	all	CIS	5	1-16	2^-29	Yes
3	all	i0 (MSB)	5	17-24	2^-31	Yes
3	all	i0 (LSB)	6	1-24		
3	all	CRC	7	1-16	2^-5	Yes
3	all	Omega (MSB)	7	17-24	2^-31	Yes
3	all	Omega (LSB)	8	1-24		
3	all	Omega Dot	9	1-24	2^-43	Yes
3	all	IODE	10	1-8	1:1	No
3	all	IDOT	10	9-22	2^-43	Yes

Satellite frequencies

GPS frequency overview

Band	Frequency	Description
L1	1575.42 MHz	Coarse-acquisition (C/A) and encrypted precision P(Y) codes, plus the L1 civilian (L1C) and military (M) codes on future Block III satellites.
L2	1227.60 MHz	P(Y) code, plus the L2C and military codes on the Block IIR-M and newer satellites.
L3	1381.05 MHz	Used for nuclear detonation (NUDET) detection.
L4	1379.913 MHz	Being studied for additional ionospheric correction.
L5	1176.45 MHz	Proposed for use as a civilian safety-of-life (SoL) signal.

All satellites broadcast at the same two frequencies, 1.57542 GHz (L1 signal) and 1.2276 GHz (L2 signal). The satellite network uses a CDMA spread-spectrum technique where the low-bitrate message data is encoded with a high-rate pseudo-random (PRN) sequence that is different for each satellite. The receiver must be aware of the PRN codes for each satellite to reconstruct the actual message data. The C/A code, for civilian use, transmits data at 1.023 million chips per second, whereas the P code, for U.S. military use, transmits at 10.23 million chips per second. The actual internal reference of the satellites is 10.22999999543 MHz to compensate for relativistic effects[74] [75] that make observers on Earth perceive a different time reference with respect to the transmitters in orbit.

The L1 carrier is modulated by both the C/A and P codes, while the L2 carrier is only modulated by the P code.[76] The P code can be encrypted as a so-called P(Y) code that is only available to military equipment with a proper decryption key. Both the C/A and P(Y) codes impart the precise time-of-day to the user.

The L3 signal at a frequency of 1.38105 GHz is used to transmit data from the satellites to ground stations. This data is used by the United States Nuclear Detonation (NUDET) Detection System (USNDS) to detect, locate, and report nuclear detonations (NUDETs) in the Earth's atmosphere and near space.[77] One usage is the enforcement of nuclear test ban treaties.

The L4 band at 1.379913 GHz is being studied for additional ionospheric correction.

The L5 frequency band at 1.17645 GHz was added in the process of GPS modernization. This frequency falls into an internationally protected range for aeronautical navigation, promising little or no interference under all circumstances. The first Block IIF satellite that would provide this signal is set to be launched in 2009.[78] The L5 consists of two carrier components that are in phase quadrature with each other. Each carrier component is bi-phase shift key (BPSK) modulated by a separate bit train. "L5, the third civil GPS signal, will eventually support safety-of-life applications for aviation and provide improved availability and accuracy."[79]

A conditional waiver has recently been granted to LightSquared to operate a terrestrial broadband service near the L1 band. Although LightSquared had applied for a license to operate in the 1525 to 1559 band as early as 2003 and it was put out for public comment, the FCC asked LightSquared to form a study group with the GPS community to test GPS receivers and identify issue that might arise due to the larger signal power from the LightSquared terrestrial network. The GPS community had not objected to the LightSquared (formerly MSV and SkyTerra) applications until November 2010, when LightSquared applied for a modification to its Ancillary Terrestrial Component (ATC) authorization. This filing (SAT-MOD-20101118-00239) amounted to a request to run several orders of magnitude more power in the same frequency band for terrestrial base stations, essentially repurposing what was supposed to be a "quiet neighborhood" for signals from space as the equivalent of a cellular network. Testing in the first half of 2011 has demonstrated that the impact of the lower 10 MHz of spectrum is minimal to GPS devices (less than 1% of the total GPS devices are affected). The upper 10 MHz intended for use by LightSquared may have some impact on GPS devices. There is some concern that this will seriously degrade the GPS signal for many consumer uses.[80] [81] Aviation Week magazine reports that the latest testing (June 2011) confirms "significant jamming" of GPS by LightSquared's system.[82]

Demodulation and decoding

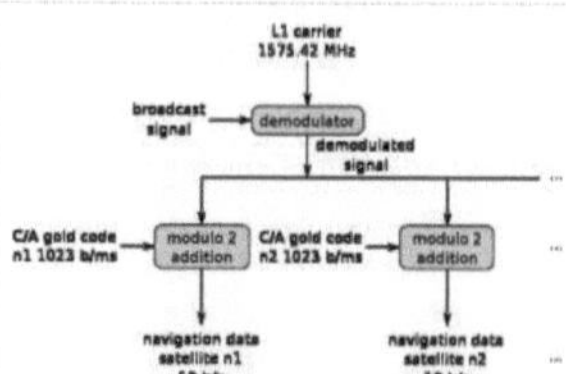

Demodulating and Decoding GPS Satellite Signals using the Coarse/Acquisition Gold code.

Because all of the satellite signals are modulated onto the same L1 carrier frequency, the signals must be separated after demodulation. This is done by assigning each satellite a unique binary sequence known as a Gold code. The signals are decoded after demodulation using addition of the Gold codes corresponding to the satellites monitored by the receiver.[83] [84]

If the almanac information has previously been acquired, the receiver picks the satellites to listen for by their PRNs, unique numbers in the range 1 through 32. If the almanac information is not in memory, the receiver enters a search mode until a lock is obtained on one of the satellites. To obtain a lock, it is necessary that there be an unobstructed line of sight from the receiver to the satellite. The receiver can then acquire the almanac and determine the satellites it should listen for. As it detects each satellite's signal, it identifies it by its distinct C/A code pattern. There can be a delay of up to 30 seconds before the first estimate of position because of the need to read the ephemeris data.

Processing of the navigation message enables the determination of the time of transmission and the satellite position at this time. For more information see Demodulation and Decoding, Advanced.

Navigation equations

The receiver uses messages received from satellites to determine the satellite positions and time sent. The *x, y,* and *z* components of satellite position and the time sent are designated as $[x_i, y_i, z_i, t_i]$ where the subscript *i* denotes the satellite and has the value 1, 2, ..., *n*, where When the time of message reception indicated by the on-board clock is , the true reception time is where is receiver's clock bias (i.e., clock delay). The message's transit time is . Assuming the message traveled at the speed of light, , the distance traveled is . Knowing the distance from receiver to satellite and the satellite's position implies that the receiver is on the surface of a sphere centered at the satellite's position with radius equal to this distance. Thus the receiver is at or near the intersection of the surfaces of the spheres if it receives signals from more than one satellite. In the ideal case of no errors, the receiver is at the intersection of the surfaces of the spheres.

The clock error or bias, *b*, is the amount that the receiver's clock is off. The receiver has four unknowns, the three components of GPS receiver position and the clock bias [*x, y, z, b*]. The equations of the sphere surfaces are given by:

or in terms of *pseudoranges*, , as

.

These equations can be solved by algebraic or numerical methods.

Bancroft's method

Bancroft's method involves an algebraic as opposed to numerical method and can be used for the case of four or more satellites.[85] [86] Bancroft's method provides one or two solutions for the four unknowns. However when there are two solutions, only one of these two solutions will be a near earth sensible solution. When there are four satellites, we use the inverse of the B matrix in section 2 of.[86] If there are more than four satellites then we use the pseudoinverse of the B matrix since in this case the B matrix is no longer square.

Trilateration

The receiver can use trilateration[87] [88] and one dimensional numerical root finding.[89] Satellite position and pseudorange determines a sphere centered on the satellite with radius equal to the pseudorange. Trilateration is used to estimate receiver position based on the intersection of three sphere surfaces so determined. In the usual case of two intersections of three sphere surfaces, the point nearest the surface of the sphere corresponding to the fourth satellite is chosen. Let *d* denote the signed distance from the current estimate of receiver position to the sphere around the fourth satellite. The notation, *d(correction)* denotes this as a function of the clock correction. The problem is to determine the correction such that *d(correction)* = 0. This is the familiar problem of finding the zeroes of a one dimensional non-linear function of a scalar variable. Iterative numerical methods, such as those found in the chapter on root finding in Numerical Recipes can solve this type of problem.[89]

Multidimensional Newton-Raphson calculations

Alternatively, multidimensional root finding methods such as the Newton-Raphson method can be used.[89] The approach is to linearize around an approximate solution, say from iteration *k*, then solve the linear equations derived from the quadratic equations above to obtain . Although there is no guarantee that the method always converges due to the fact that multidimensional roots cannot be bounded, when a neighborhood containing a solution is known as is usually the case for GPS, it is quite likely that a solution will be found.[89] It has been shown[90] that results are comparable in accuracy to those of Bancroft's method.

Additional methods for more than four satellites

When more than four satellites are available, the calculation can use the four best or more than four, considering number of channels, processing capability, and geometric dilution of precision (GDOP). Using more than four is an over-determined system of equations with no unique solution, which must be solved by least-squares or a similar technique.[86] If all visible satellites are used, the results are as good as or better than using the four best. Errors can be estimated through the residuals. With each combination of four or more satellites, a GDOP factor can be calculated, based on the relative sky directions of the satellites used.[91] As more satellites are picked up, pseudoranges from various 4-way combinations can be processed to add more estimates to the location and clock offset. The receiver then takes the weighted average of these positions and clock offsets. After the final location and time are calculated, the location is expressed in a specific coordinate system such as latitude and longitude, using the WGS 84 geodetic datum or a country-specific system.[92]

Error sources and analysis

GPS error analysis which is found in Error analysis for the Global Positioning System is an important aspect for determining what errors and their magnitude are to be expected. GPS errors are affected by geometric dilution of precision and depend on signal arrival time errors, numerical errors, atmospherics effects, ephemeris errors, multipath errors and other effects. Variability in solar radiation pressure[93] has an indirect effect on GPS accuracy due to its effect on ephemeris errors.

Accuracy enhancement and surveying

Augmentation

Integrating external information into the calculation process can materially improve accuracy. Such augmentation systems are generally named or described based on how the information arrives. Some systems transmit additional error information (such as clock drift, ephemera, or ionospheric delay), others characterize prior errors, while a third group provides additional navigational or vehicle information.

Examples of augmentation systems include the Wide Area Augmentation System (WAAS), European Geostationary Navigation Overlay Service (EGNOS), Differential GPS, Inertial Navigation Systems (INS) and Assisted GPS.

Precise monitoring

Accuracy can be improved through precise monitoring and measurement of existing GPS signals in additional or alternate ways.

The largest remaining error is usually the unpredictable delay through the ionosphere. The spacecraft broadcast ionospheric model parameters, but errors remain. This is one reason GPS spacecraft transmit on at least two frequencies, L1 and L2. Ionospheric delay is a well-defined function of frequency and the total electron content (TEC) along the path, so measuring the arrival time difference between the frequencies determines TEC and thus the precise ionospheric delay at each frequency.

Military receivers can decode the P(Y)-code transmitted on both L1 and L2. Without decryption keys, it is still possible to use a *codeless* technique to compare the P(Y) codes on L1 and L2 to gain much of the same error information. However, this technique is slow, so it is currently available only on specialized surveying equipment. In the future, additional civilian codes are expected to be transmitted on the L2 and L5 frequencies (see GPS modernization). Then all users will be able to perform dual-frequency measurements and directly compute ionospheric delay errors.

A second form of precise monitoring is called *Carrier-Phase Enhancement* (CPGPS). This corrects the error that arises because the pulse transition of the PRN is not instantaneous, and thus the correlation (satellite-receiver sequence matching) operation is imperfect. CPGPS uses the L1 carrier wave, which has a period of , which is about one-thousandth of the C/A Gold code bit period of , to act as an additional clock signal and resolve the uncertainty. The phase difference error in the normal GPS amounts to 2–3 metres ( ) of ambiguity. CPGPS working to within 1% of perfect transition reduces this error to 3 centimetres (in) of ambiguity. By eliminating this error source, CPGPS coupled with DGPS normally realizes between 20–30 centimetres ( ) of absolute accuracy.

Relative Kinematic Positioning (RKP) is a third alternative for a precise GPS-based positioning system. In this approach, determination of range signal can be resolved to a precision of less than 10 centimetres (in). This is done by resolving the number of cycles that the signal is transmitted and received by the receiver by using a combination of differential GPS (DGPS) correction data, transmitting GPS signal phase information and ambiguity resolution techniques via statistical tests—possibly with processing in real-time (real-time kinematic positioning, RTK).

Timekeeping

Leap seconds

While most clocks are synchronized to Coordinated Universal Time (UTC), the atomic clocks on the satellites are set to GPS time (GPST; see the page of United States Naval Observatory). The difference is that GPS time is not corrected to match the rotation of the Earth, so it does not contain leap seconds or other corrections that are periodically added to UTC. GPS time was set to match Coordinated Universal Time (UTC) in 1980, but has since diverged. The lack of corrections means that GPS time remains at a constant offset with International Atomic Time (TAI) (TAI − GPS = 19 seconds). Periodic corrections are performed on the on-board clocks to keep them synchronized with ground clocks.[94]

The GPS navigation message includes the difference between GPS time and UTC. As of July 2012, GPS time is 16 seconds ahead of UTC because of the leap second added to UTC June 30, 2012.[95] Receivers subtract this offset from GPS time to calculate UTC and specific timezone values. New GPS units may not show the correct UTC time until after receiving the UTC offset message. The GPS-UTC offset field can accommodate 255 leap seconds (eight bits) that, given the current period of the Earth's rotation (with one leap second introduced approximately every 18 months), should be sufficient to last until approximately the year 2300.

Accuracy

GPS time is theoretically accurate to about 14 nanoseconds.[96] However, most receivers lose accuracy in the interpretation of the signals and are only accurate to 100 nanoseconds.[97] [98]

Format

As opposed to the year, month, and day format of the Gregorian calendar, the GPS date is expressed as a week number and a seconds-into-week number. The week number is transmitted as a ten-bit field in the C/A and P(Y) navigation messages, and so it becomes zero again every 1,024 weeks (19.6 years). GPS week zero started at 00:00:00 UTC (00:00:19 TAI) on January 6, 1980, and the week number became zero again for the first time at 23:59:47 UTC on August 21, 1999 (00:00:19 TAI on August 22, 1999). To determine the current Gregorian date, a GPS receiver must be provided with the approximate date (to within 3,584 days) to correctly translate the GPS date signal. To address this concern the modernized GPS navigation message uses a 13-bit field that only repeats every 8,192 weeks (157 years), thus lasting until the year 2137 (157 years after GPS week zero).

Carrier phase tracking (surveying)

Another method that is used in surveying applications is carrier phase tracking. The period of the carrier frequency multiplied by the speed of light gives the wavelength, which is about 0.19 meters for the L1 carrier. Accuracy within 1% of wavelength in detecting the leading edge reduces this component of pseudorange error to as little as 2 millimeters. This compares to 3 meters for the C/A code and 0.3 meters for the P code.

However, 2 millimeter accuracy requires measuring the total phase—the number of waves multiplied by the wavelength plus the fractional wavelength, which requires specially equipped receivers. This method has many surveying applications.

Triple differencing followed by numerical root finding, and a mathematical technique called least squares can estimate the position of one receiver given the position of another. First, compute the difference between satellites, then between receivers, and finally between epochs. Other orders of taking differences are equally valid. Detailed discussion of the errors is omitted.

The satellite carrier total phase can be measured with ambiguity as to the number of cycles. Let denote the phase of the carrier of satellite *j* measured by receiver *i* at time . This notation shows the meaning of the subscripts *i, j,* and *k.* The receiver (*r*), satellite (*s*), and time (*t*) come in alphabetical order as arguments of and to balance readability and conciseness, let be a concise abbreviation. Also we define three functions, :, which return differences between receivers, satellites, and time points, respectively. Each function has variables with three subscripts as its arguments. These three functions are defined below. If is a function of the three integer arguments, *i, j,* and *k* then it is a valid argument for the functions, :, with the values defined as

,

, and

.

Also if are valid arguments for the three functions and *a* and *b* are constants then is a valid argument with values defined as

,

, and

.

Receiver clock errors can be approximately eliminated by differencing the phases measured from satellite 1 with that from satellite 2 at the same epoch.[99] This difference is designated as

Double differencing[100] computes the difference of receiver 1's satellite difference from that of receiver 2. This approximately eliminates satellite clock errors. This double difference is:

Triple differencing[101] subtracts the receiver difference from time 1 from that of time 2. This eliminates the ambiguity associated with the integral number of wavelengths in carrier phase provided this ambiguity does not change with time. Thus the triple difference result eliminates practically all clock bias errors and the integer ambiguity. Atmospheric delay and satellite ephemeris errors have been significantly reduced. This triple difference is:

Triple difference results can be used to estimate unknown variables. For example if the position of receiver 1 is known but the position of receiver 2 unknown, it may be possible to estimate the position of receiver 2 using numerical root finding and least squares. Triple difference results for three independent time pairs quite possibly will be sufficient to solve for receiver 2's three position components. This may require the use of a numerical procedure.[102] [103] An approximation of receiver 2's position is required to use such a numerical method. This initial value can probably be provided from the navigation message and the intersection of sphere surfaces. Such a reasonable estimate can be key to successful multidimensional root finding. Iterating from three time pairs and a fairly good initial value produces one observed triple difference result for receiver 2's position. Processing additional time pairs can improve accuracy, overdetermining the answer with multiple solutions. Least squares can estimate an overdetermined system. Least squares determines the position of receiver 2 which best fits the observed triple difference results for receiver 2 positions under the criterion of minimizing the sum of the squares.

Regulatory spectrum issues concerning GPS receivers

In the United States, GPS receivers are regulated under the Federal Communications Commission's (FCC) Part 15 rules. As indicated in the manuals of GPS-enabled devices sold in the United States, as a Part 15 device, it "must accept any interference received, including interference that may cause undesired operation."[104] With respect to GPS devices in particular, the FCC states that GPS receiver manufacturers, "must use receivers that reasonably discriminate against reception of signals outside their allocated spectrum."[105]

The spectrum allocated for GPS L1 use by the FCC is 1559 to 1610 MHz.[106] Since 1996, the FCC has authorized licensed use of the spectrum neighboring the GPS band of 1525 to 1559 MHz to the Virginia company

LightSquared. On March 1, 2001, the FCC received an application from LightSquared's predecessor, Motient Services to use their allocated frequencies for an integrated satellite-terrestrial service.[107] In 2002, the U.S. GPS Industry Council came to an out-of-band-emissions (OOBE) agreement with LightSquared to prevent transmissions from LightSquared's ground-based stations from emitting transmissions into the neighboring GPS band of 1559 to 1610 MHz.[108] In 2004, the FCC adopted the OOBE agreement in its authorization for LightSquared to deploy a ground-based network that used its allocated frequencies of 1525 to 1559 MHz.[109] This authorization was reviewed and approved by the U.S. Interdepartment Radio Advisory Committee, which includes the U.S. Department of Agriculture, U.S. Air Force, U.S. Army, U.S. Coast Guard, Federal Aviation Administration, National Aeronautics and Space Administration, Interior, and U.S. Department of Transportation.[110]

In January 2011, the FCC conditionally authorized LightSquared's wholesale customers, such as Best Buy, Sharp, and C Spire, to be able to only purchase an integrated satellite-ground-based service from LightSquared and re-sell that integrated service on devices that are equipped to only use the ground-based signal using LightSquared's allocated frequencies of 1525 to 1559 MHz.[111] In December 2010, GPS receiver manufacturers expressed concerns to the FCC that LightSquared's signal would interfere with GPS receiver devices[112] although the FCC's policy considerations leading up to the January 2011 order did not pertain to any proposed changes to the maximum number of ground-based LightSquared stations or the maximum power at which these stations could operate. The January 2011 order makes final authorization contingent upon studies of GPS interference issues carried out by a LightSquared led working group along with GPS industry and Federal agency participation.

GPS receiver manufacturers design GPS receivers to use spectrum beyond the GPS-allocated band. In some cases, GPS receivers are designed to use up to 400 MHz of spectrum in either direction of the L1 frequency of 1575.42 MHz.[113] However, as regulated under the FCC's Part 15 rules, GPS receivers are not warranted protection from signals outside GPS-allocated spectrum.[105]

The FCC adopted rules in February 2003 that allowed Mobile Satellite Service (MSS) licensees such as LightSquared to construct ground-based towers in their licensed spectrum to "promote more efficient use of terrestrial wireless spectrum."[114] In July 2010, the FCC stated that it expected LightSquared to use its authority to offer an integrated satellite-terrestrial service to "provide mobile broadband services similar to those provided by terrestrial mobile providers and enhance competition in the mobile broadband sector."[115] However, GPS receiver manufacturers have argued that LightSquared's licensed spectrum of 1525 to 1559 MHz was never envisioned as being used for high-speed wireless broadband although there is no regulatory or legal backing of this claim.[116] To build public support of efforts to reverse the 2004 FCC authorization of LightSquared's network, GPS receiver manufacturer Trimble Navigation Ltd. formed the "Coalition To Save Our GPS."[117]

The FCC and LightSquared have each made public commitments to solve the GPS interference issue before the network is allowed to operate.[118] [119] However, according to Chris Dancy of the Aircraft Owners and Pilots Association, airline pilots with the type of systems that would be affected "may go off course and not even realize it."[120] The problems could also affect the Federal Aviation Administration upgrade to the air traffic control system, United States Defense Department guidance, and local emergency services including 911.[120]

On February 14, 2012, the U.S. Federal Communications Commission (FCC) moved to bar LightSquared's planned national broadband network after being informed by the National Telecommunications and Information Administration (NTIA), "the federal agency that coordinates spectrum uses for the military and other federal government entities," that "there is no practical way to mitigate potential interference at this time."[121] [122] LightSquared is challenging the FCC's action.

Other systems

Other satellite navigation systems in use or various states of development include:

- GLONASS – Russia's global navigation system. Fully operational worldwide.
- Galileo – a global system being developed by the European Union and other partner countries, planned to be operational by 2014 (and fully deployed by 2019)
- Beidou – People's Republic of China's regional system, currently limited to Asia and the West Pacific[125]
- COMPASS – People's Republic of China's global system, planned to be operational by 2020[126] [127]
- IRNSS – India's regional navigation system, planned to be operational by 2012, covering India and Northern Indian Ocean[128]
- QZSS – Japanese regional system covering Asia and Oceania

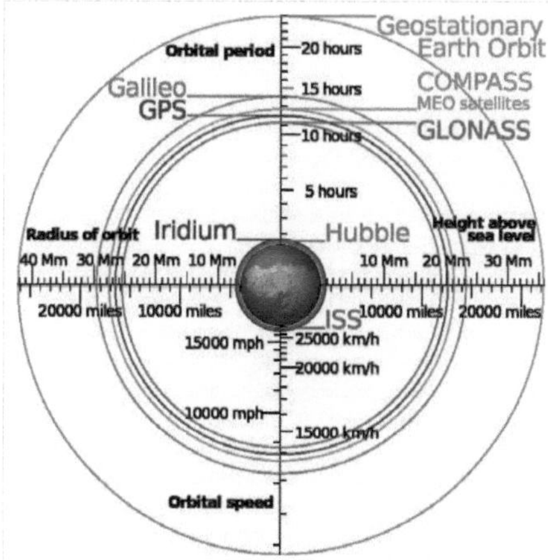

Comparison of **GPS**, **GLONASS**, **Galileo** and **Compass (medium earth orbit)** satellite navigation system orbits with the International Space Station, Hubble Space Telescope and Iridium constellation orbits, Geostationary Earth Orbit, and the nominal size of the Earth.[123] The Moon's orbit is around 9 times larger (in radius and length) than geostationary orbit.[124]

See also

- GLONASS
- GPS/INS
- GPS navigation software
- High Sensitivity GPS
- Indoor positioning system
- Local positioning system
- Military invention
- Mobile phone tracking
- Navigation paradox
- S-GPS

Notes

[1] National Research Council (U.S.). Committee on the Future of the Global Positioning System; National Academy of Public Administration (1995). *The global positioning system: a shared national asset: recommendations for technical improvements and enhancements* (http://books.google.com/books?id=FAHk65slfY4C). National Academies Press. p. 16. ISBN 0-309-05283-1. ., Chapter 1, p. 16 (http://books.google.com/books?id=FAHk65slfY4C&pg=PA16)

[2] "Factsheets : GPS Advanced Control Segment (OCX)" (http://www.losangeles.af.mil/library/factsheets/factsheet.asp?id=18676). Losangeles.af.mil. October 25, 2011. . Retrieved November 6, 2011.

[3] *Astronautica Acta II*, 25 (1956)

[4] "GPS and Relativity" (http://www.astronomy.ohio-state.edu/~pogge/Ast162/Unit5/gps.html). Astronomy.ohio-state.edu. . Retrieved November 6, 2011.

[5] Guier, William H.; Weiffenbach, George C. (1997). "Genesis of Satellite Navigation" (http://www.jhuapl.edu/techdigest/td/td1901/guier.pdf). *John Hopkins APL Technical Digest* **19** (1): 178–181. .

[6] Steven Johnson (2010), *Where good ideas come from, the natural history of innovation*, New York: Riverhead Books

[7] Jerry Proc. "Omega" (http://www.jproc.ca/hyperbolic/omega.html). Jproc.ca. . Retrieved December 8, 2009.

[8] "Why Did the Department of Defense Develop GPS?" (http://web.archive.org/web/20071018151253/http://www.trimble.com/gps/whygps.shtml#0). Trimble Navigation Ltd. Archived from the original (http://www.trimble.com/gps/whygps.shtml#0) on October 18, 2007. . Retrieved January 13, 2010.
[9] "Charting a Course Toward Global Navigation" (http://www.aero.org/publications/crosslink/summer2002/01.html). The Aerospace Corporation. . Retrieved January 14, 2010.
[10] "A Guide to the Global Positioning System (GPS) — GPS Timeline" (http://support.radioshack.com/support_tutorials/gps/gps_tmline.htm). Radio Shack. . Retrieved January 14, 2010.
[11] "SECOR Chronology" (http://www.astronautix.com/craft/secor.htm). *Mark Wade's Encyclopedia Astronautica*. . Retrieved January 19, 2010.
[12] Michael Russell Rip, James M. Hasik (2002). *The Precision Revolution: GPS and the Future of Aerial Warfare* (http://books.google.com/?id=mB9W3H90KDUC). Naval Institute Press. p. 65. ISBN 1-55750-973-5. . Retrieved January 14, 2010.
[13] "ICAO Completes Fact-Finding Investigation" (http://www.icao.int/cgi/goto_m.pl?icao/en/trivia/kal_flight_007.htm). International Civil Aviation Organization. . Retrieved September 15, 2008.
[14] "United States Updates Global Positioning System Technology" (http://www.america.gov/xarchives/display.html?p=washfile-english&y=2006&m=February&x=20060203125928lcnirellep0.5061609). America.gov (http://www.america.gov/). February 3, 2006. .
[15] "GPS & Selective Availability Q&A" (http://ngs.woc.noaa.gov/FGCS/info/sans_SA/docs/GPS_SA_Event_QAs.pdf). (http://www.noaa.gov/). . Retrieved May 28, 2010.
[16] "GPS Modernization Fact Sheet" (http://www.pnt.gov/public/docs/2006/modernization.pdf). U.S. Air Force. .
[17] GPS Wing Reaches GPS III IBR Milestone (http://www.insidegnss.com/node/918) in InsideGNSS November 10, 2008
[18] "GPS almanacs" (http://www.navcen.uscg.gov/?pageName=gpsAlmanacs). Navcen.uscg.gov. . Retrieved October 15, 2010.
[19] Dietrich Schroeer, Mirco Elena (2000). *Technology Transfer* (http://books.google.com/?id=I7JRAAAAMAAJ). Ashgate. p. 80. ISBN 0-7546-2045-X. . Retrieved May 25, 2008.
[20] Michael Russell Rip, James M. Hasik (2002). *The Precision Revolution: GPS and the Future of Aerial Warfare* (http://books.google.com/?id=_wpUAAAAMAAJ). Naval Institute Press. ISBN 1-55750-973-5. . Retrieved May 25, 2008.
[21] "AF Space Command Chronology" (http://www.afspc.af.mil/heritage/chronology.asp). USAF Space Command. . Retrieved June 20, 2011.
[22] "FactSheet: 2nd Space Operations Squadron" (http://www.schriever.af.mil/library/factsheets/factsheet.asp?id=4045). USAF Space Command. . Retrieved June 20, 2011.
[23] The Global Positioning System: Assessing National Policies (http://www.rand.org/pubs/monograph_reports/MR614/MR614.appb.pdf), p.245. RAND corporation
[24] "USNO NAVSTAR Global Positioning System" (http://tycho.usno.navy.mil/gpsinfo.html). U.S. Naval Observatory. . Retrieved January 7, 2011.
[25] National Archives and Records Administration. U.S. Global Positioning System Policy (http://clinton4.nara.gov/textonly/WH/EOP/OSTP/html/gps-factsheet.html). March 29, 1996.
[26] "National Executive Committee for Space-Based Positioning, Navigation, and Timing" (http://pnt.gov/). Pnt.gov. . Retrieved October 15, 2010.
[27] "Assisted-GPS Test Calls for 3G WCDMA Networks" (http://www.3g.co.uk/PR/November2004/8641.htm). 3g.co.uk. November 10, 2004. . Retrieved November 24, 2010.
[28] This story was written by 010907 (September 17, 2007). "losangeles.af.mil" (http://www.losangeles.af.mil/news/story.asp?id=123068412). losangeles.af.mil. . Retrieved October 15, 2010.
[29] Johnson, Bobbie (May 19, 2009). "GPS system 'close to breakdown'" (http://www.guardian.co.uk/technology/2009/may/19/gps-close-to-breakdown). *The Guardian* (London). . Retrieved December 8, 2009.
[30] Coursey, David (May 21, 2009). "Air Force Responds to GPS Outage Concerns" (http://abcnews.go.com/Technology/AheadoftheCurve/story?id=7647002&page=1). ABC News. . Retrieved May 22, 2009.
[31] "Air Force GPS Problem: Glitch Shows How Much U.S. Military Relies On GPS" (http://www.huffingtonpost.com/2010/06/01/air-force-gps-problem-gli_n_595727.html). Huffingtonpost.comm. June 1, 2010. . Retrieved October 15, 2010.
[32] "Contract Award for Next Generation GPS Control Segment Announced" (http://www.losangeles.af.mil/news/story_print.asp?id=123192234). . Retrieved December 14, 2012.
[33] "United States Naval Observatory (USNO) GPS Constellation Status" (ftp://tycho.usno.navy.mil/pub/gps/gpstd.txt). . Retrieved October 13, 2009.
[34] United States Naval Observatory. GPS Constellation Status (ftp://tycho.usno.navy.mil/pub/gps/gpsb2.txt). Retrieved December 20, 2008.
[35] "United Launch Alliance GPS IIF-2" (http://www.ulalaunch.com/site/pages/News.shtml#/73/). United Launch Alliance. . Retrieved July 16, 2011.
[36] "United Launch Alliance GPS IIF-3" (http://www.ulalaunch.com/site/pages/News.shtml#/121/). United Launch Alliance. . Retrieved October 8, 2012.
[37] United States Naval Research Laboratory. National Medal of Technology for GPS (http://www.eurekalert.org/pub_releases/2005-11/nrl-par112205.php). November 21, 2005

[38] "Space Technology Hall of Fame, Inducted Technology: Global Positioning System (GPS)" (http://www.spacetechhalloffame.org/inductees_1998_Global_Positioning_System.html). .

[39] This can be seen from the fact that three sphere surfaces typically intersect at two points as shown in trilateration and for a fourth sphere surface to intersect the other three, it would have to go through one of the points at which the other three intersect. This is a special case not the general situation. Also because of the fact that two intersections are typical for three sphere surfaces, we can say that three satellites are inadequate.

[40] Georg zur Bonsen, Daniel Ammann, Michael Ammann, Etienne Favey, Pascal Flammant (April 1, 2005). "Continuous Navigation Combining GPS with Sensor-Based Dead Reckoning" (http://web.archive.org/web/20061111202317/http://www.gpsworld.com/gpsworld/article/articleDetail.jsp?id=154870&pageID=6). GPS World. Archived from the original (http://www.gpsworld.com/gpsworld/article/articleDetail.jsp?id=154870&pageID=6) on November 11, 2006. .

[41] "NAVSTAR GPS User Equipment Introduction" (http://www.navcen.uscg.gov/pubs/gps/gpsuser/gpsuser.pdf) (PDF). United States Government. . Chapter 7

[42] "GPS Support Notes" (http://web.archive.org/web/20090327051208/http://www.navmanwireless.com/uploads/EK/C8/EKC8zb1lTsNwDqWcqLQxiQ/Support_Notes_GPS_OperatingParameters.pdf) (PDF). January 19, 2007. Archived from the original (http://www.navmanwireless.com/uploads/EK/C8/EKC8zb1lTsNwDqWcqLQxiQ/Support_Notes_GPS_OperatingParameters.pdf) on March 27, 2009. . Retrieved November 10, 2008.

[43] John Pike. "GPS III Operational Control Segment (OCX)" (http://www.globalsecurity.org/space/systems/gps_3-ocx.htm). Globalsecurity.org. . Retrieved December 8, 2009.

[44] "Global Positioning System" (http://www.gps.gov/systems/gps/index.html). Gps.gov. . Retrieved June 26, 2010.

[45] Daly, P.. "Navstar GPS and GLONASS: global satellite navigation systems" (http://ieeexplore.ieee.org/iel1/2219/7072/00285510.pdf?arnumber=285510). IEEE. .

[46] Dana, Peter H. (August 8, 1996). "GPS Orbital Planes" (http://www.colorado.edu/geography/gcraft/notes/gps/gif/oplanes.gif) (GIF). .

[47] What the Global Positioning System Tells Us about Relativity (http://metaresearch.org/cosmology/gps-relativity.asp). Retrieved January 2, 2007.

[48] GPS Overview from the NAVSTAR Joint Program Office (http://www.losangeles.af.mil/library/factsheets/factsheet.asp?id=5325). Retrieved December 15, 2006.

[49] (http://www.gmat.unsw.edu.au/snap/gps/gps_survey/chap2/222sats.htm). Retrieved October 27, 2011

[50] "USCG Navcen: GPS Frequently Asked Questions" (http://www.navcen.uscg.gov/?pageName=gpsFaq). . Retrieved January 31, 2007.

[51] Agnew, D.C. and Larson, K.M. (2007). "Finding the repeat times of the GPS constellation". *GPS Solutions* (Springer) **11** (1): 71–76. doi:10.1007/s10291-006-0038-4. This article from author's web site (http://spot.colorado.edu/~kristine/gpsrep.pdf), with minor correction.

[52] "CURRENT GPS CONSTELLATION" (http://tycho.usno.navy.mil/gpscurr.html). U.S. Naval Observatory. .

[53] Massatt, Paul; Wayne Brady (Summer 2002). "Optimizing performance through constellation management" (http://www.aero.org/publications/crosslink/summer2002/index.html). *Crosslink*: 17–21. .

[54] United States Coast Guard General GPS News 9–9–05 (http://igs.bkg.bund.de/root_ftp/IGS/mail/igsmail/year2005/5209)

[55] USNO NAVSTAR Global Positioning System (http://tycho.usno.navy.mil/gpsinfo.html). Retrieved May 14, 2006.

[56] "GPS III Operational Control Segment (OCX)" (http://www.globalsecurity.org/space/systems/gps_3-ocx.htm). GlobalSecurity.org. .

[57] "The USA's GPS-III Satellites" (http://www.defenseindustrydaily.com/The-USAs-GPS-III-Satellites-04900/). Defense Industry Daily. October 13, 2011. .

[58] "GPS Completes Next Generation Operational Control System PDR" (http://www.comspacewatch.com/news/viewpr.html?pid=34625). Air Force Space Command News Service. September 14, 2011. .

[59] Though there are many receiver manufacturers, they almost all use one of the chipsets produced for this purpose. An example: "GPS Receiver Chip Performance Survey" (http://gpstekreviews.com/2007/04/14/gps-receiver-chip-performance-survey/). GPS Technology Reviews. .

[60] "Publications and Standards from the National Marine Electronics Association (NMEA)" (http://www.nmea.org/pub/index.html). National Marine Electronics Association. . Retrieved June 27, 2008.

[61] "Common View GPS Time Transfer" (http://tf.nist.gov/time/commonviewgps.htm). nist.gov. . Retrieved 2011-Jul-23.

[62] "Spotlight GPS pet locator" (http://www.spotlightgps.com/). Spotlightgps.com. . Retrieved October 15, 2010.

[63] Arms Control Association. Missile Technology Control Regime (http://www.armscontrol.org/documents/mtcr). Retrieved May 17, 2006.

[64] "Commanders Digital Assistant explanation and photo" (http://web.archive.org/web/20071201034857/http://peosoldier.army.mil/factsheets/SWAR_LW_DBCS.pdf). Web.archive.org. December 1, 2007. Archived from the original (http://peosoldier.army.mil/factsheets/SWAR_LW_DBCS.pdf) on December 1, 2007. . Retrieved November 6, 2011.

[65] "Latest version Commanders Digital Assistant" (http://peosoldier.army.mil/factsheets/SWAR_LW_CDA.pdf) (PDF). . Retrieved October 13, 2009.

[66] "Soldier Digital Assistant explanation and photo" (http://web.archive.org/web/20080610092154/http://www.army-technology.com/contractors/computers/lago/lago6.html). Web.archive.org. June 10, 2008. Archived from the original (http://www.army-technology.com/contractors/computers/lago/lago6.html) on June 10, 2008. . Retrieved November 6, 2011.

[67] Sinha, Vandana (July 24, 2003). "Commanders and Soldiers' GPS-receivers" (http://gcn.com/articles/2003/07/24/soldiers-take-digital-assistants-to-war.aspx). Gcn.com. . Retrieved October 13, 2009.

[68] "XM982 Excalibur Precision Guided Extended Range Artillery Projectile" (http://www.globalsecurity.org/military/systems/munitions/m982-155.htm). GlobalSecurity.org. May 29, 2007. . Retrieved September 26, 2007.

[69] Sandia National Laboratory's Nonproliferation programs and arms control technology (http://www.sandia.gov/LabNews/LN03-07-03/LA2003/la03/arms_story.htm).

[70] Dr. Dennis D. McCrady. "The GPS Burst Detector W-Sensor" (http://www.osti.gov/bridge/servlets/purl/10176800-S2tU7w/native/10176800.pdf). Sandia National Laboratories. .

[71] "Satellite message format" (http://gpsinformation.net/gpssignal.htm). Gpsinformation.net. . Retrieved October 15, 2010.

[72] Peter H. Dana. "GPS Week Number Rollover Issues" (http://www.colorado.edu/geography/gcraft/notes/gps/gpseow.htm). .

[73] "Interface Specification IS-GPS-200, Revision D: Navstar GPS Space Segment/Navigation User Interfaces" (http://www.losangeles.af.mil/shared/media/document/AFD-070803-059.pdf) (PDF). Navstar GPS Joint Program Office. . Page 103.

[74] Misra, Pratap; Enge, Per (2006). *Global Positioning System. Signals, Measurements and Performance* (http://books.google.com/books?id=pv5MAQAAIAAJ) (2nd ed.). Ganga-Jamuna Press. p. 115. ISBN 0-9709544-1-7. .

[75] Borre, Kai; M. Akos, Dennis; Bertelsen, Nicolaj; Rinder, Peter; Jensen, Søren Holdt (2007). *A Software-Defined GPS and Galileo Receiver. A single-Frequency Approach* (http://books.google.com/books?id=x2g6XTEkb8oC). Springer. p. 18. ISBN 0-8176-4390-7. .

[76] How GPS works. (http://www.kowoma.de/en/gps/signals.htm) Konowa.de (2005).

[77] TextGenerator Version 2.0. "United States Nuclear Detonation Detection System (USNDS)" (http://www.fas.org/spp/military/program/nssrm/initiatives/usnds.htm). Fas.org. . Retrieved November 6, 2011.

[78] First GPS IIF Satellite Undergoes Environmental Testing (http://mg.gpsworld.com/gpsmg/Military+&+Government+News/First-GPS-IIF-Satellite-Undergoes-Environmental-Te/ArticleStandard/Article/detail/470408?contextCategoryId=33824). GPS World. November 5, 2007.

[79] "Air Force Successfully Transmits an L5 Signal From GPS IIR-20(M) Satellite" (http://www.losangeles.af.mil/news/story.asp?storyID=123144001). LA AFB News Release. . Retrieved June 20, 2011.

[80] "Federal Communications Commission Presented Evidence of GPS Signal Interference" (http://www.gpsworld.com/gnss-system/news/data-shows-disastrous-gps-jamming-fcc-approved-broadcaster-11029?utm_source=GPS&utm_medium=email&utm_campaign=Navigate_01_31_2011&utm_content=data-shows-disastrous-gps-jamming-fcc-approved-broadcaster-11029). GPS World. . Retrieved November 6, 2011.

[81] "Coalition to Save Our GPS" (http://www.saveourgps.org/studies-reports.aspx). Saveourgps.org. . Retrieved November 6, 2011.

[82] "LightSquared Tests Confirm GPS Jamming" (http://www.aviationweek.com/aw/generic/story.jsp?id=news/awx/2011/06/09/awx_06_09_2011_p0-334122.xml&headline=LightSquared Tests Confirm GPS Jamming&channel=busav). Aviation Week. . Retrieved June 20, 2011.

[83] "GPS Almanacs, NANUS, and Ops Advisories (including archives)" (http://www.navcen.uscg.gov/?pageName=gpsAlmanacs). *GPS Almanac Information*. United States Coast Guard. . Retrieved September 9, 2009.

[84] "George, M., Hamid, M., and Miller A. Gold Code Generators in Virtex Devices (http://web.archive.org/web/20071122063244/http://www.xilinx.com/support/documentation/application_notes/xapp217.pdf) PDF (126 KB)

[85] Bancroft, S. (January 1985). "An Algebraic Solution of the GPS Equations" (http://ieeexplore.ieee.org/xpls/abs_all.jsp?arnumber=4104017). *IEEE Transactions on Aerospace and Electronic Systems* **AES-21**: 56–59. Bibcode 1985ITAES..21...56B. doi:10.1109/TAES.1985.310538. .

[86] "Global Positioning Systems" (http://web.archive.org/web/20110719232148/http://www.macalester.edu/~halverson/math36/GPS.pdf) (PDF). . Retrieved October 15, 2010.

[87] Position Determination with GPS. (http://www.kowoma.de/en/gps/positioning.htm) Konowa.de (2005).

[88] How GPS Receivers Work (http://electronics.howstuffworks.com/gadgets/travel/gps2.htm) at How Stuff Works

[89] Press, Flannery, Tekolsky, and Vetterling 1986, *Numerical Recipes, The Art of Scientific Computing* (Cambridge University Press).

[90] Kumar, B. Hari; Reddy, K. Chennakesava; Namassivaya, N. (November 1, 2008). "Determination of GPS receiver position using Multivariate Newton-Raphson Technique for over specified cases" (http://www.thefreelibrary.com/Determination+of+GPS+receiver+position+using+Multivariate...-a0216041321). *International Journal of Applied Engineering Research* **13** (11): 1457–1460. .

[91] Dana, Peter H.. "Geometric Dilution of Precision (GDOP) and Visibility" (http://www.colorado.edu/geography/gcraft/notes/gps/gps.html#Gdop). University of Colorado at Boulder. . Retrieved July 7, 2008.

[92] Peter H. Dana. "Receiver Position, Velocity, and Time" (http://www.colorado.edu/geography/gcraft/notes/gps/gps.html#PosVelTime). University of Colorado at Boulder. . Retrieved July 7, 2008.

[93] *IPN Progress Report* 42-159 (2004) (http://ipnpr.jpl.nasa.gov/progress_report/42-159/159I.pdf)

[94] "NAVSTAR GPS User Equipment Introduction" (http://www.navcen.uscg.gov/pubs/gps/gpsuser/gpsuser.pdf) (PDF). . Section 1.2.2

[95] "Notice Advisory to Navstar Users (NANU) 2012034" (https://gps.afspc.af.mil/gps/archive/2012/nanus/2012034.nnu). GPS Operations Center. May 30, 2012. . Retrieved July 2, 2012.

[96] David W. Allan (1997). *The Science of Timekeeping* (http://www.allanstime.com/Publications/DWA/Science_Timekeeping/TheScienceOfTimekeeping.pdf). Hewlett Packard. Archived (http://www.webcitation.org/6BMeuPXJs) from the original on October 12, 2012. .

[97] *The Role of GPS in Precise Time and Frequency Dissemination* (http://ilrs.gsfc.nasa.gov/docs/timing/gpsrole.pdf). GPSworld. July/August 1990. . Retrieved October 12, 2012.

[98] "GPS time accurate to 100 nanoseconds" (http://www.atomic-clock.galleon.eu.com/support/gps-time-accuracy.html). Galleon. . Retrieved October 12, 2012.

[99] "Between-Satellite Differencing" (http://www.gmat.unsw.edu.au/snap/gps/gps_survey/chap6/633.htm). Gmat.unsw.edu.au. . Retrieved October 15, 2010.

[100] "Double differencing" (http://www.gmat.unsw.edu.au/snap/gps/gps_survey/chap6/635.htm). Gmat.unsw.edu.au. . Retrieved October 15, 2010.

[101] "Triple differencing" (http://www.gmat.unsw.edu.au/snap/gps/gps_survey/chap6/636.htm). Gmat.unsw.edu.au. . Retrieved October 15, 2010.

[102] chapter on root finding and nonlinear sets of equations

[103] *Preview of Root Finding* (http://books.google.com/?id=UQW_VL2H56IC&pg=PA959&lpg=PA959&dq="Numerical+Analysis"+multidimension++root+finding#PPA442,M1). Books.google.com. 2007. ISBN 978-0-521-88068-8. . Retrieved October 15, 2010.

[104] "2011 John Deere StarFire 3000 Operator Manual" (http://stellarsupport.deere.com/en_US/support/pdf/om/en/ompfp11008_sf3000.pdf). John Deere. . Retrieved November 13, 2011.

[105] "Federal Communications Commission Report and Order In the Matter of Fixed and Mobile Services in the Mobile Satellite Service Bands at 1525-1559 MHz and 1626.5-1660.5 MHz" (http://hraunfoss.fcc.gov/edocs_public/attachmatch/FCC-11-57A1.pdf). FCC.gov. April 6, 2011. . Retrieved December 13, 2011.

[106] "Federal Communications Commission Table of Frequency Allocations" (http://transition.fcc.gov/oet/spectrum/table/fcctable.pdf). FCC.gov. November 18, 2011. . Retrieved December 13, 2011.

[107] "FCC Docket File Number: SATASG2001030200017, "Mobile Satellite Ventures LLC Application for Assignment and Modification of Licenses and for Authority to Launch and Operate a Next-Generation Mobile Satellite System"" (http://licensing.fcc.gov/cgi-bin/ws.exe/prod/ib/forms/reports/related_filing.hts?f_key=200647&f_number=SATASG2001030200017). FCC.gov. March 1, 2001. p. 9. .

[108] "U.S. GPS Industry Council Petition to the FCC to adopt OOBE limits jointly proposed by MSV and the Industry Council" (http://fjallfoss.fcc.gov/ecfs/document/view?id=6515082621). FCC.gov. September 4, 2003. . Retrieved December 13, 2011.

[109] "Federal Communications Commission Order and Authorization Granting Authority" (http://hraunfoss.fcc.gov/edocs_public/attachmatch/DA-04-3553A1.pdf). FCC.gov. November 8, 2004. . Retrieved December 13, 2011.

[110] "Statement of Julius P. Knapp, Chief, Office of Engineering and Technology, Federal Communications Commission" (http://www.gps.gov/congress/hearings/2011-09-HASC/knapp.pdf). gps.gov. September 15, 2011. p. 3. . Retrieved December 13, 2011.

[111] "FCC Order, Granted LightSquared Subsidiary LLC, a Mobile Satellite Service licensee in the L-Band, a conditional waiver of the Ancillary Terrestrial Component "integrated service" rule" (http://hraunfoss.fcc.gov/edocs_public/attachmatch/DA-11-133A1.pdf). *Federal Communications Commission*. FCC.Gov. January 26, 2011. . Retrieved December 13, 2011.

[112] "Data Shows Disastrous GPS Jamming from FCC-Approved Broadcaster" (http://www.gpsworld.com/gnss-system/news/data-shows-disastrous-gps-jamming-fcc-approved-broadcaster-11029). gpsworld.com. February 1, 2011. . Retrieved February 10, 2011.

[113] "Javad Ashjaee GPS World webinar" (http://www.gpsworld.com/gnss-system/news/javad-ashjaee-discuss-javad-gnss-lightsquared-tech-december-8-webinar-12337). gpsworld.com. December 8, 2011. . Retrieved December 13, 2011.

[114] "FCC Order permitting mobile satellite services providers to provide an ancillary terrestrial component (ATC) to their satellite systems" (http://hraunfoss.fcc.gov/edocs_public/attachmatch/FCC-03-15A1.pdf). *Federal Communications Commission*. FCC.gov. February 10, 2003. . Retrieved December 13, 2011.

[115] "Federal Communications Commission Fixed and Mobile Services in the Mobile Satellite Service" (http://www.federalregister.gov/articles/2010/08/16/2010-19824/fixed-and-mobile-services-in-the-mobile-satellite-service#p-31). *Federal Communications Commission*. FCC.gov. July 15, 2010. . Retrieved December 13, 2011.

[116] http://saveourgps.org/pdf/SIS_DOD_Response_Statement_08122011.pdf

[117] "Coalition to Save Our GPS" (http://saveourgps.org/). Saveourgps.org. . Retrieved November 6, 2011.

[118] Jeff Carlisle (June 23, 2011). "Testimony of Jeff Carlisle, LightSquared Executive Vice President of Regulatory Affairs and Public Policy to U.S. House Subcommittee on Aviation and Subcommittee on Coast Guard and Maritime Transportation" (http://ssv.cachefly.net/lightsquared/wp-content/uploads/2011/06/LSQ-Testimony-Package.pdf). . Retrieved December 13, 2011.

[119] Julius Genachowski (May 31, 2011). "FCC Chairman Genachowski Letter to Senator Charles Grassley" (http://www.lightsquared.com/documents/FCC Julius Genachowski letter to Senator Grassley - May 31, 2011.pdf). . Retrieved December 13, 2011.

[120] Tessler, Joelle (April 7, 2011). "Internet network may jam GPS in cars, jets" (http://www.thesunnews.com/2011/04/07/2085752/internet-network-may-jam-gps-in.html). *The Sun News*. . Retrieved April 7, 2011.

[121] FCC press release "" Spokesperson Statement on NTIA Letter--LightSquared and GPS (http://www.fcc.gov/document/spokesperson-statement-ntia-letter-lightsquared-and-gps)" February 14, 2012

[122] Paul Riegler, FBT. " FCC Bars LightSquared Broadband Network Plan (http://www.frequentbusinesstraveler.com/2012/02/fcc-bars-lightsquared-broadband-network-plan/)." February 14, 2012. Retrieved February 14, 2012.

[123] Orbital periods and speeds are calculated using the relations $4\pi^2R^3 = T^2GM$ and $V^2R = GM$, where R = radius of orbit in metres, T = orbital period in seconds, V = orbital speed in m/s, G = gravitational constant ≈ 6.673×10^{-11} Nm²/kg², M = mass of Earth ≈ 5.98×10^{24} kg.

[124] Approximately 8.6 times when the moon is nearest (363 104 km ÷ 42 164 km) to 9.6 times when the moon is farthest (405 696 km ÷ 42 164 km).

[125] Beidou coverage

[126] "Beidou satellite navigation system to cover whole world in 2020" (http://eng.chinamil.com.cn/news-channels/china-military-news/2010-05/20/content_4222569.htm). Eng.chinamil.com.cn.. Retrieved October 15, 2010.

[127] Levin, Dan (March 23, 2009). "Chinese Square Off With Europe in Space" (http://www.nytimes.com/2009/03/23/technology/23iht-galileo23.html?_r=1&scp=1&sq=chinese europe galileo&st=cse). *The New York Times* (China).. Retrieved November 6, 2011.

[128] "ASM, News on GIS, GNSS, spatial information, remote sensing, mapping and surveying technologies for Asia" (http://www.asmmag.com/news/india-to-launch-1st-irnss-satellite-by-december). Asmmag.com.. Retrieved October 13, 2009.

References

- "NAVSTAR GPS User Equipment Introduction" (http://www.navcen.uscg.gov/pubs/gps/gpsuser/gpsuser.pdf) (PDF). United States Coast Guard. September 1996.

Further reading

- Parkinson; Spilker (1996). *The global positioning system* (http://books.google.com/?id=lvI1a5J_4ewC). American Institute of Aeronautics and Astronautics. ISBN 978-1-56347-106-3.
- Jaizki Mendizabal; Roc Berenguer; Juan Melendez (2009). *GPS and Galileo* (http://books.google.com/?id=t1lBTH42mOcC&printsec=frontcover&dq=GPS+and+GALILEO#v=onepage&q&f=false). McGraw Hill. ISBN 978-0-07-159869-9.
- Nathaniel Bowditch (2002). *The American Practical Navigator – Chapter 11 Satellite Navigation*. United States government.
- Global Positioning System (http://ocw.mit.edu/courses/earth-atmospheric-and-planetary-sciences/12-540-principles-of-the-global-positioning-system-spring-2012/) Open Courseware from MIT, 2012

External links

- Schrieve Air Force Base - GPS Operations Center (http://www.schriever.af.mil/GPS/) Responsible for operation of the Global Positioning System
- Global Positioning System (http://www.dmoz.org/Science/Earth_Sciences/Geomatics/Global_Positioning_System/) at the Open Directory Project
- FAA GPS FAQ (http://www.faa.gov/about/office_org/headquarters_offices/ato/service_units/techops/navservices/gnss/faq/gps/)
- GPS.gov (http://www.gps.gov/)—General public education website created by the U.S. Government
- National Space-Based PNT Executive Committee (http://pnt.gov/) established 2004 to oversee management of GPS and GPS augmentations at national level
- USCG Navigation Center (http://www.navcen.uscg.gov/?pageName=GPS)—Status of the GPS constellation, government policy, and links to other references; includes satellite almanac data.
- The GPS Program Office (GPS Wing) (http://gps.losangeles.af.mil/)—Responsible for designing and acquiring the system on behalf of the United States Government.
- U.S. Army Corps of Engineers manual: NAVSTAR HTML (http://web.archive.org/web/20080822132227/http://www.usace.army.mil/publications/eng-manuals/em1110-1-1003/toc.htm) and PDF (22.6 MB, 328 pages) (http://web.archive.org/web/20080625111519/http://www.usace.army.mil/publications/eng-manuals/em1110-1-1003/entire.pdf)
- National Geodetic Survey (http://www.ngs.noaa.gov/orbits/) Orbits for the Global Positioning System satellites in the Global Navigation Satellite System
- GPS SPS Performance Standard (http://pnt.gov/public/docs/2008/spsps2008.pdf)—The official Standard Positioning Service specification
- GPS PPS Performance Standard (http://www.ion.org/search/view_abstract.cfm?jp=p&idno=7517)—The official Precise Positioning Service specification

- GPS and GLONASS Simulation (http://rhp.detmich.com/gps.html) (Java applet) Simulation and graphical depiction of space vehicle motion including computation of dilution of precision (DOP)

Root_cause_analysis

Root cause analysis (**RCA**) is a method of problem solving that tries to identify the root causes of faults or problems that cause operating events.

RCA practice tries to solve problems by attempting to identify and correct the root causes of events, as opposed to simply addressing their symptoms. By focusing correction on root causes, problem recurrence can be prevented. RCFA (Root Cause Failure Analysis) recognizes that complete prevention of recurrence by one corrective action is not always possible.

Nevertheless, in the U.S. nuclear power industry the NRC requires that "In the case of significant conditions adverse to quality, the measures shall assure that the cause of the condition is determined and corrective action taken to prevent repetition." [10CFR50, Appendix B, Criterion XVI, Sentence 2)] In practice more than one "cause" is allowed and more than one corrective action is not forbidden.

Conversely, there may be several effective measures (methods) that address the root causes of a problem. Thus, RCA is often considered to be an iterative process, and is frequently viewed as a tool of continuous improvement.

RCA is typically used as a reactive method of identifying event(s) causes, revealing problems and solving them. Analysis is done *after* an event has occurred. Insights in RCA may make it useful as a pro-active method. In that event, RCA can be used to *forecast* or predict probable events even *before* they occur. While one follows the other, RCA is a completely separate process to Incident Management.

Root cause analysis is not a single, sharply defined methodology; there are many different tools, processes, and philosophies for performing RCA. However, several very-broadly defined approaches or "schools" can be identified by their basic approach or field of origin: safety-based, production-based, process-based, failure-based, and systems-based.

- Safety-based RCA descends from the fields of accident analysis and occupational safety and health.
- Production-based RCA has its origins in the field of quality control for industrial manufacturing.
- Process-based RCA is basically a follow-on to production-based RCA, but with a scope that has been expanded to include business processes.
- Failure-based RCA is rooted in the practice of failure analysis as employed in engineering and maintenance.
- Systems-based RCA has emerged as an amalgamation of the preceding schools, along with ideas taken from fields such as change management, risk management, and systems analysis.

Despite the different approaches among the various schools of root cause analysis, there are some common principles. It is also possible to define several general processes for performing RCA.

General principles of root cause analysis

1. The primary aim of RCA is to identify the factors that resulted in the nature, the magnitude, the location, and the timing of the harmful outcomes (consequences) of one or more past events in order to identify what behaviours, actions, inactions, or conditions need to be changed to prevent recurrence of similar harmful outcomes and to identify the lessons to be learned to promote the achievement of better consequences. ("Success" is defined as the near-certain prevention of recurrence.)
2. To be effective, RCA must be performed systematically, usually as part of an investigation, with conclusions and root causes that are identified backed up by documented evidence. Usually a team effort is required.

3. There may be more than one root cause for an event or a problem, the difficult part is demonstrating the persistence and sustaining the effort required to determine them.
4. The purpose of identifying all solutions to a problem is to prevent recurrence at lowest cost in the simplest way. If there are alternatives that are equally effective, then the simplest or lowest cost approach is preferred.
5. Root causes identified depend on the way in which the problem or event is defined. Effective problem statements and event descriptions (as failures, for example) are helpful, or even required.
6. To be effective, the analysis should establish a sequence of events or timeline to understand the relationships between contributory (causal) factors, root cause(s) and the defined problem or event to prevent in the future.
7. Root cause analysis can help to transform a reactive culture (that reacts to problems) into a forward-looking culture that solves problems before they occur or escalate. More importantly, it reduces the frequency of problems occurring over time within the environment where the RCA process is used.
8. RCA is a threat to many cultures and environments. Threats to cultures often meet with resistance. There may be other forms of management support required to achieve RCA effectiveness and success. For example, a "non-punitive" policy toward problem identifiers may be required.

General process for performing and documenting an RCA-based Corrective Action

Notice that RCA (in steps 3, 4 and 5) forms the most critical part of successful corrective action, because it directs the corrective action at the true root cause of the problem. The root cause is secondary to the goal of prevention, but without knowing the root cause, it is not possible to determine what an effective corrective action for the defined problem will be.

1. Define the problem or describe the event factually. Include the qualitative and quantitative attributes (properties) of the harmful outcomes. This usually includes specifying the natures, the magnitudes, the locations, and the timing of events.
2. Gather data and evidence, classifying it along a timeline of events to the final failure or crisis. For every behavior, condition, action, and inaction specify in the "timeline" what should have been done when it differs from what was done.
3. Ask "why" and identify the causes associated with each step in the sequence towards the defined problem or event. "Why" is taken to mean "What were the factors that directly resulted in the effect?"
4. Classify causes into causal factors that relate to an event in the sequence and root causes, that if eliminated, can be agreed to have interrupted that step of the sequence chain.
5. Identify all other harmful factors that have equal or better claim to be called "root causes." If there are multiple root causes, which is often the case, reveal those clearly for later optimum selection.
6. Identify corrective action(s) that will with certainty prevent recurrence of each harmful effect, including outcomes and factors. Check that each corrective action would, if pre-implemented before the event, have reduced or prevented specific harmful effects.
7. Identify solutions that, when effective, and with consensus agreement of the group, prevent recurrence with reasonable certainty , are within the institution's control, meet its goals and objectives and do not cause or introduce other new, unforeseen problems.
8. Implement the recommended root cause correction(s).
9. Ensure effectiveness by observing the implemented recommendation solutions.
10. Identify other methodologies for problem solving and problem avoidance that may be useful.
11. Identify and address the other instances of each harmful outcome and harmful factor.

See also

- Failure mode and effects analysis
- Fault tree analysis
- Forensic engineering
- Fix it twice
- Eight Disciplines Problem Solving
- Multiple regression and multivariate linear regression

References

Time

The flow of sand in an hourglass can be used to keep track of elapsed time. It also concretely represents the present as being between the past and the future.

Time is a dimension in which events can be ordered from the past through the present into the future,[1] [2] [3] [4] [5] [6] and also the measure of durations of events and the intervals between them.[3] [7] [8] Time has long been a major subject of study in religion, philosophy, and science, but defining it in a manner applicable to all fields without circularity has consistently eluded scholars.[3] [7] [8] [9] [10] [11] Nevertheless, diverse fields such as business, industry, sports, the sciences, music, dance, and the live theater all incorporate some notion of time into their respective measuring systems.[12] [13] [14] Some simple, relatively uncontroversial definitions of time include "time is what clocks measure"[7] [15] and "time is what keeps everything from happening at once".[16] [17] [18] [19]

Two contrasting viewpoints on time divide many prominent philosophers. One view is that time is part of the fundamental structure of the universe—a dimension independent of events, in which events occur in sequence. Sir Isaac Newton subscribed to this realist view, and hence it is sometimes referred to as Newtonian time.[20] [21] The opposing view is that *time* does not refer to any kind of "container" that events and objects "move through", nor to any entity that "flows", but that it is instead part of a fundamental intellectual structure (together with space and number) within which humans sequence and compare events. This second view, in the tradition of Gottfried Leibniz[15] and Immanuel Kant,[22] [23] holds that *time* is neither an event nor a thing, and thus is not itself measurable nor can it be travelled.

Time is one of the seven fundamental physical quantities in the International System of Units. Time is used to define other quantities – such as velocity — so defining time in terms of such quantities would result in circularity of definition.[24] An operational definition of time, wherein one says that observing a certain number of repetitions of one or another standard cyclical event (such as the passage of a free-swinging pendulum) constitutes one standard unit such as the second, is highly useful in the conduct of both advanced experiments and everyday affairs of life. The operational definition leaves aside the question whether there is something called time, apart from the counting activity just mentioned, that flows and that can be measured. Investigations of a single continuum called spacetime bring questions about space into questions about time, questions that have their roots in the works of early students of natural philosophy.

Furthermore, it may be that there is a subjective component to time, but whether or not time itself is "felt", as a sensation or an experience, has never been settled.[3] [7] [8] [25] [26]

Temporal measurement has occupied scientists and technologists, and was a prime motivation in navigation and astronomy. Periodic events and periodic motion have long served as standards for units of time. Examples include the apparent motion of the sun across the sky, the phases of the moon, the swing of a pendulum, and the beat of a heart. Currently, the international unit of time, the second, is defined in terms of radiation emitted by caesium atoms (see below). Time is also of significant social importance, having economic value ("time is money") as well as personal value, due to an awareness of the limited time in each day and in human life spans.

Temporal measurement

Temporal measurement, or chronometry, takes two distinct period forms: the calendar, a mathematical abstraction for calculating extensive periods of time,[27] and the clock, a physical mechanism that counts the ongoing passage of time. In day-to-day life, the clock is consulted for periods less than a day, the calendar, for periods longer than a day. Increasingly, personal electronic devices display both calendars and clocks simultaneously. The number (as on a clock dial or calendar) that marks the occurrence of a specified event as to hour or date is obtained by counting from a fiducial epoch – a central reference point.

History of the calendar

Artifacts from the Palaeolithic suggest that the moon was used to reckon time as early as 6,000 years ago.[28] Lunar calendars were among the first to appear, either 12 or 13 lunar months (either 354 or 384 days). Without intercalation to add days or months to some years, seasons quickly drift in a calendar based solely on twelve lunar months. Lunisolar calendars have a thirteenth month added to some years to make up for the difference between a full year (now known to be about 365.24 days) and a year of just twelve lunar months. The numbers twelve and thirteen came to feature prominently in many cultures, at least partly due to this relationship of months to years.

The reforms of Julius Caesar in 45 BC put the Roman world on a solar calendar. This Julian calendar was faulty in that its intercalation still allowed the astronomical solstices and equinoxes to advance against it by about 11 minutes per year. Pope Gregory XIII introduced a correction in 1582; the Gregorian calendar was only slowly adopted by different nations over a period of centuries, but is today by far the one in most common use around the world.

History of time measurement devices

A large variety of devices have been invented to measure time. The study of these devices is called horology.

Horizontal sundial in Taganrog.

An Egyptian device that dates to c.1500 BC, similar in shape to a bent T-square, measured the passage of time from the shadow cast by its crossbar on a nonlinear rule. The T was orientated eastward in the mornings. At noon, the device was turned around so that it could cast its shadow in the evening direction.[29]

A sundial uses a gnomon to cast a shadow on a set of markings calibrated to the hour. The position of the shadow marks the hour in local time.

The most precise timekeeping device of the ancient world was the water clock, or *clepsydra*, one of which was found in the tomb of Egyptian pharaoh Amenhotep I (1525–1504 BC). They could be used to measure the hours even at night, but required manual upkeep to replenish the flow of water. The Greeks and the people from Chaldea (southern

Iraq) regularly maintained timekeeping records as an essential part of their astronomical observations. Arab inventors and engineers in particular made improvements on the use of water clocks up to the Middle Ages.[30] In the 11th century, Chinese inventors and engineers invented the first mechanical clocks driven by an escapement mechanism.

A contemporary quartz watch

The hourglass uses the flow of sand to measure the flow of time. They were used in navigation. Ferdinand Magellan used 18 glasses on each ship for his circumnavigation of the globe (1522).[31] Incense sticks and candles were, and are, commonly used to measure time in temples and churches across the globe. Waterclocks, and later, mechanical clocks, were used to mark the events of the abbeys and monasteries of the Middle Ages. Richard of Wallingford (1292–1336), abbot of St. Alban's abbey, famously built a mechanical clock as an astronomical orrery about 1330.[32] [33] Great advances in accurate time-keeping were made by Galileo Galilei and especially Christiaan Huygens with the invention of pendulum driven clocks.

The English word clock probably comes from the Middle Dutch word *klocke*—which, in turn, derives from the mediaeval Latin word *clocca*, which ultimately derives from Celtic and is cognate with French, Latin, and German words that mean bell. The passage of the hours at sea were marked by bells, and denoted the time (see ship's bells). The hours were marked by bells in abbeys as well as at sea.

Clocks can range from watches, to more exotic varieties such as the Clock of the Long Now. They can be driven by a variety of means, including gravity, springs, and various forms of electrical power, and regulated by a variety of means such as a pendulum.

A chronometer is a portable timekeeper that meets certain precision standards. Initially, the term was used to refer to the marine chronometer, a timepiece used to determine longitude by means of celestial navigation, a precision firstly achieved by John Harrison. More recently, the term has also been applied to the chronometer watch, a wristwatch that meets precision standards set by the Swiss agency COSC.

Chip-scale atomic clocks, such as this one unveiled in 2004, are expected to greatly improve GPS location.[34]

The 555 timer IC is an integrated circuit (chip) used in a variety of timer, pulse generator and oscillator applications.

The most accurate timekeeping devices are atomic clocks, which are accurate to seconds in many millions of years,[35] and are used to calibrate other clocks and timekeeping instruments. Atomic clocks use the spin property of atoms as their basis, and since 1967, the International System of Measurements bases its unit of time, the second, on the properties of caesium atoms. SI defines the second as 9,192,631,770 cycles of the radiation that corresponds to the transition between two electron spin energy levels of the ground state of the ^{133}Cs atom.

Today, the Global Positioning System in coordination with the Network Time Protocol can be used to synchronize timekeeping systems across the globe.

In medieval philosophical writings, the **atom** was a unit of time referred to as the smallest possible division of time. The earliest known occurrence in English is in Byrhtferth's *Enchiridion* (a science text) of 1010–1012,[36] where it was defined as 1/564 of a *momentum* (1½ minutes),[37] and thus equal to 15/94 of a second. It was used in the *computus*, the process of calculating the date of Easter.

As of May 2010, the smallest time interval uncertainty in direct measurements is on the order of 12 attoseconds (1.2×10^{-17} seconds), about 3.7×10^{26} Planck times.[38]

Definitions and standards

Units of time

Unit	Size	Notes
yoctosecond	10^{-24} s	
zeptosecond	10^{-21} s	
attosecond	10^{-18} s	shortest time uncertainty in present measurements
femtosecond	10^{-15} s	pulse time of ultrafast lasers (100 as = 0.1 fs)
picosecond	10^{-12} s	
nanosecond	10^{-9} s	time for molecules to fluoresce
microsecond	10^{-6} s	
millisecond	0.001 s	
second	1 s	SI base unit
minute	60 seconds	
hour	60 minutes	
day	24 hours	
week	7 days	Also called *sennight*
fortnight	14 days	2 weeks
lunar month	27.2–29.5 days	Various definitions exist.
lunation	29.53 days	One lunar phase cycle.

month	28–31 days	
quarter	3 months	
year	12 months	
common year	365 days	52 weeks + 1-day
leap year	366 days	52 weeks + 2 days
tropical year	365.24219 days[39]	average
Gregorian year	365.2425 days[40]	average
Julian year	365.25 days	
Olympiad	4-year cycle	
lustrum	5 years	Also called *pentad*
decade	10 years	
Indiction	15-year cycle	
generation	17–35 years	approximate
jubilee	50 years	
century	100 years	
millennium	1,000 years	
exasecond	10^{18} s	roughly 32 billion years, more than twice the age of the universe on current estimates
cosmological decade	varies	10 times the length of the previous cosmological decade, with CĐ 1 beginning either 10 seconds or 10 years after the Big Bang, depending on the definition.

The SI base unit for time is the SI second. From the second, larger units such as the minute, hour and day are defined, though they are "non-SI" units because they do not use the decimal system, and also because of the occasional need for a leap second. They are, however, officially accepted for use *with* the International System. There are no fixed ratios between seconds and months or years as months and years have significant variations in length.[41]

The official SI definition of the second is as follows:[41] [42]

> The second is the duration of 9,192,631,770 periods of the radiation corresponding to the transition between the two hyperfine levels of the ground state of the caesium 133 atom.

At its 1997 meeting, the CIPM affirmed that this definition refers to a caesium atom in its ground state at a temperature of 0 K.[41] Previous to 1967, the second was defined as:

> the fraction 1/31,556,925.9747 of the tropical year for 1900 January 0 at 12 hours ephemeris time.

The current definition of the second, coupled with the current definition of the metre, is based on the special theory of relativity, which affirms our space-time to be a Minkowski space.

World time

Time-keeping is so critical to the functioning of modern societies that it is coordinated at an international level. The basis for scientific time is a continuous count of seconds based on atomic clocks around the world, known as the International Atomic Time (TAI). Other scientific time standards include Terrestrial Time and Barycentric Dynamical Time.

Coordinated Universal Time (UTC) is the basis for modern civil time. Since 1 January 1972, it has been defined to follow TAI with an exact offset of an integer number of seconds, changing only when a leap second is added to keep clock time synchronized with the rotation of the Earth. In TAI and UTC systems, the duration of a second is constant, as it is defined by the unchanging transition period of the caesium atom.

Greenwich Mean Time (GMT) is an older standard, adopted starting with British railways in 1847. Using telescopes instead of atomic clocks, GMT was calibrated to the mean solar time at the Royal Observatory, Greenwich in the UK. Universal Time (UT) is the modern term for the international telescope-based system, adopted to replace "Greenwich Mean Time" in 1928 by the International Astronomical Union. Observations at the Greenwich Observatory itself ceased in 1954, though the location is still used as the basis for the coordinate system. Because the rotational period of Earth is not perfectly constant, the duration of a second would vary if calibrated to a telescope-based standard like GMT or UT—in which a second was defined as a fraction of a day or year. The terms "GMT" and "Greenwich Mean Time" are sometimes used informally to refer to UT or UTC.

The Global Positioning System also broadcasts a very precise time signal worldwide, along with instructions for converting GPS time to UTC.

Earth is split up into a number of time zones. Most time zones are exactly one hour apart, and by convention compute their local time as an offset from UTC or GMT. In many locations these offsets vary twice yearly due to daylight saving time transitions.

Time conversions

These conversions are accurate at the millisecond level for time systems involving earth rotation (UT1 & TT). Conversions between atomic time systems (TAI, GPS, and UTC) are accurate at the microsecond level.

System	Description	UT1	UTC	TT	TAI	GPS
UT1	Mean Solar Time	UT1	UTC = UT1 - DUT1	TT = UT1 + 32.184 s + LS - DUT1	TAI = UT1 - DUT1 + LS	GPS = UT1 - DUT1 + LS - 19 s
UTC	Civil Time	UT1 = UTC + DUT1	UTC	TT = UTC + 32.184 s + LS	TAI = UTC + LS	GPS = UTC + LS - 19 s
TT	Terrestrial (Ephemeris) Time	UT1 = TT - 32.184 s - LS + DUT1	UTC = TT - 32.184 s - LS	TT	TAI = TT - 32.184 s	GPS = TT - 51.184 s
TAI	Atomic Time	UT1 = TAI + DUT1 - LS	UTC = TAI - LS	TT = TAI + 32.184 s	TAI	GPS = TAI - 19 s
GPS	GPS Time	UT1 = GPS + DUT1 - LS + 19 s	UTC = GPS - LS + 19 s	TT = GPS + 51.184 s	TAI = GPS + 19 s	GPS

Definitions:

1. LS = TAI - UTC = Leap Seconds from http://maia.usno.navy.mil/ser7/tai-utc.dat
2. DUT1 = UT1 - UTC from http://maia.usno.navy.mil/ser7/ser7.dat or http://maia.usno.navy.mil/search/search.html

Sidereal time

Sidereal time is the measurement of time relative to a distant star (instead of solar time that is relative to the sun). It is used in astronomy to predict when a star will be overhead. Due to the orbit of the earth around the sun a sidereal day is 4 minutes (1/366th) less than a solar day.

Chronology

Another form of time measurement consists of studying the past. Events in the past can be ordered in a sequence (creating a chronology), and can be put into chronological groups (periodization). One of the most important systems of periodization is geologic time, which is a system of periodizing the events that shaped the Earth and its life. Chronology, periodization, and interpretation of the past are together known as the study of history.

Religion

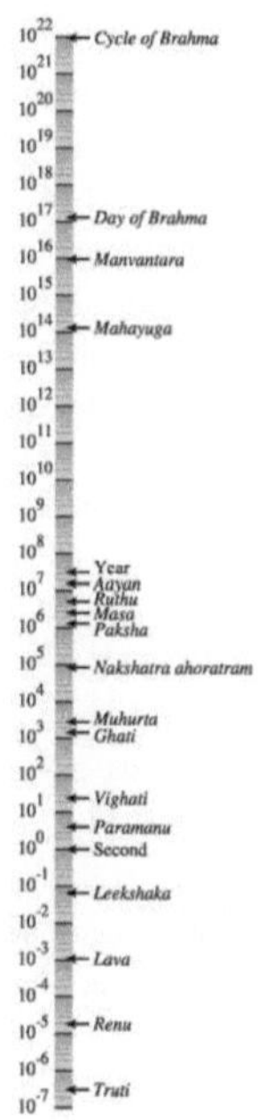

Hindu units of time shown logarithmically

Further information: Time and fate deities

Linear and cyclical time

Ancient cultures such as Incan, Mayan, Hopi, and other Native American Tribes, plus the Babylonians, Ancient Greeks, Hinduism, Buddhism, Jainism, and others have a concept of a wheel of time, that regards time as cyclical and quantic consisting of repeating ages that happen to every being of the Universe between birth and extinction.

In general, the Judaeo-Christian concept, based on the Bible, is that time is linear, beginning with the act of creation by God. The general Christian view is that time will end with the end of the world.

In the Old Testament book Ecclesiastes, traditionally ascribed to Solomon (970–928 BC), time (as the Hebrew word עדן, זמן *ʿiddan(time) zĕman(season)* is often translated) was traditionally regarded as a medium for the passage of predestined events. (Another word, زمان" זמן" *zman*, was current as meaning *time fit for an event*, and is used as the modern Arabic, Persian, and Hebrew equivalent to the English word "time".)

> There is an appointed time (*zman*) for everything. And there is a time ('êth) for every event under heaven–
> A time (*'êth*) to give birth, and a time to die; A time to plant, and a time to uproot what is planted.
> A time to kill, and a time to heal; A time to tear down, and a time to build up.
> A time to weep, and a time to laugh; A time to mourn, and a time to dance.
> A time to throw stones, and a time to gather stones; A time to embrace, and a time to shun embracing.
> A time to search, and a time to give up as lost; A time to keep, and a time to throw away.
> A time to tear apart, and a time to sew together; A time to be silent, and a time to speak.
> A time to love, and a time to hate; A time for war, and a time for peace. – Ecclesiastes 3:1–8

Time in Greek mythology

The Greek language denotes two distinct principles, Chronos and Kairos. The former refers to numeric, or chronological, time. The latter, literally "the right or opportune moment", relates specifically to metaphysical or Divine time. In theology, Kairos is qualitative, as opposed to quantitative.

In Greek mythology, Chronos (Ancient Greek: Χρόνος) is identified as the Personification of Time. His name in Greek means "time" and is alternatively spelled Chronus (Latin spelling) or Khronos. Chronos is usually portrayed as an old, wise man with a long, gray beard, such as "Father Time". Some English words whose etymological root is khronos/chronos include *chronology*, *chronometer*, *chronic*, *anachronism*, *synchronize*, and *chronicle*.

Philosophy

Two distinct viewpoints on time divide many prominent philosophers. One view is that time is part of the fundamental structure of the universe, a dimension in which events occur in sequence. Sir Isaac Newton subscribed to this realist view, and hence it is sometimes referred to as Newtonian time.[21] An opposing view is that *time* does not refer to any kind of actually existing dimension that events and objects "move through", nor to any entity that "flows", but that it is instead an intellectual concept (together with space and number) that enables humans to sequence and compare events[43] This second view, in the tradition of Gottfried Leibniz[15] and Immanuel Kant,[22][23] holds that space and time "do not exist in and of themselves, but ... are the product of the way we represent things", because we can know objects only as they appear to us.

The *Vedas*, the earliest texts on Indian philosophy and Hindu philosophy dating back to the late 2nd millennium BC, describe ancient Hindu cosmology, in which the universe goes through repeated cycles of creation, destruction and rebirth, with each cycle lasting 4320 million years.[44] Ancient Greek philosophers, including Parmenides and Heraclitus, wrote essays on the nature of time.[45] Plato, in the *Timaeus*, identified time with the period of motion of the heavenly bodies. Aristotle, in Book IV of his *Physica* defined time as the number of change with respect to before and after.

In Book 11 of his *Confessions*, St. Augustine of Hippo ruminates on the nature of time, asking, "What then is time? If no one asks me, I know: if I wish to explain it to one that asketh, I know not." He begins to define time by what it is not rather than what it is,[46] an approach similar to that taken in other negative definitions. However, Augustine ends up calling time a “distention” of the mind (Confessions 11.26) by which we simultaneously grasp the past in memory, the present by attention, and the future by expectation.

In contrast to ancient Greek philosophers who believed that the universe had an infinite past with no beginning, medieval philosophers and theologians developed the concept of the universe having a finite past with a beginning. This view is shared by Abrahamic faiths as they believe time started by creation, therefore the only thing being infinite is God and everything else, including time, is finite.

Isaac Newton believed in absolute space and absolute time; Leibniz believed that time and space are relational.[47] The differences between Leibniz's and Newton's interpretations came to a head in the famous Leibniz-Clarke Correspondence.

> "Time is not an empirical concept. For neither co-existence nor succession would be perceived by us, if the representation of time did not exist as a foundation *a priori*. Without this presupposition we could not represent to ourselves that things exist together at one and the same time, or at different times, that is, contemporaneously, or in succession."
>
> Immanuel Kant, *Critique of Pure Reason* (1781), trans. Vasilis Politis (London: Dent., 1991), p.54.

Immanuel Kant, in the *Critique of Pure Reason*, described time as an *a priori* intuition that allows us (together with the other *a priori* intuition, space) to comprehend sense experience.[48] With Kant, neither space nor time are conceived as substances, but rather both are elements of a systematic mental framework that necessarily structures the experiences of any rational agent, or observing subject. Kant thought of time as a fundamental part of an abstract conceptual framework, together with space and number, within which we sequence events, quantify their duration, and compare the motions of objects. In this view, *time* does not refer to any kind of entity that "flows," that objects "move through," or that is a "container" for events. Spatial measurements are used to quantify the extent of and distances between objects, and temporal measurements are used to quantify the durations of and between events. (See Ontology).

Henri Bergson believed that time was neither a real homogeneous medium nor a mental construct, but possesses what he referred to as *Duration*. Duration, in Bergson's view, was creativity and memory as an essential component of reality.[49]

According to Martin Heidegger we do not exist inside time, *we are time*. Hence, the relationship to the past is a present awareness of *having been*, which allows the past to exist in the present. The relationship to the future is the state of anticipating a potential possibility, task, or engagement. It is related to the human propensity for caring and being concerned, which causes "being ahead of oneself" when thinking of a pending occurrence. Therefore, this concern for a potential occurrence also allows the future to exist in the present. The present becomes an experience, which is qualitative instead of quantitative. Heidegger seems to think this is the way that a linear relationship with time, or temporal existence, is broken or transcended.[50] We are not stuck in sequential time. We are able to remember the past and project into the future - we have a kind of random access to our representation of temporal existence --- we can, in our thoughts, step out of (ecstasis) sequential time.[51]

Time as "unreal"

In 5th century BC Greece, Antiphon the Sophist, in a fragment preserved from his chief work *On Truth*, held that: *"Time is not a reality (hypostasis), but a concept (noêma) or a measure (metron)."* Parmenides went further, maintaining that time, motion, and change were illusions, leading to the paradoxes of his follower Zeno.[52] Time as an illusion is also a common theme in Buddhist thought.[53] [54]

J. M. E. McTaggart's 1908 *The Unreality of Time* argues that, since every event has the characteristic of being both present and not present (i.e., future or past), that time is a self-contradictory idea (see also The flow of time).

These arguments often center around what it means for something to be *unreal*. Modern physicists generally believe that time is as *real* as space—though others, such as Julian Barbour in his book *The End of Time*, argue that quantum equations of the universe take their true form when expressed in the timeless realm containing every possible *now* or momentary configuration of the universe, called 'platonia' by Barbour.[55] (See also: Eternalism (philosophy of time))

Physical definition

From the age of Newton to Einstein's profound reinterpretation of the physical concepts associated with time and space, time was considered *absolute* and to flow *equably* (to use Newton's words) for all observers.[56] Non-relativistic classical mechanics is based on this Newtonian idea of time.

Einstein, in his special theory of relativity,[57] postulated the constancy and finiteness of the speed of light for all observers. He showed that this postulate, together with a reasonable definition for what it means for two events to be simultaneous, requires that distances appear compressed and time intervals appear lengthened for events associated with objects in motion relative to an inertial observer.

The theory of special relativity finds a convenient formulation in Minkowski spacetime, a mathematical structure that combines three dimensions of space with a single dimension of time. In this formalism, distances in space can be measured by how long light takes to travel that distance, e.g., a light-year is a measure of distance, and a meter is now defined in terms of how far light travels in a certain amount of time. Two events in Minkowski spacetime are separated by an *invariant interval*, which can be either space-like, light-like, or time-like. Events that are time-like cannot be simultaneous in any frame of reference, there must be a temporal component (and possibly a spatial one) to their separation. Events that are space-like could be simultaneous in some frame of reference, and there is no frame of reference in which they do not have a spatial separation. People travelling at different velocities between two events measure different spatial and temporal separations between the events, but the *invariant interval* is constant and independent of velocity.

Classical mechanics

In non-relativistic classical mechanics, Newton's concept of "relative, apparent, and common time" can be used in the formulation of a prescription for the synchronization of clocks. Events seen by two different observers in motion relative to each other produce a mathematical concept of time that works sufficiently well for describing the everyday phenomena of most people's experience. In the late nineteenth century, physicists encountered problems with the classical understanding of time, in connection with the behaviour of electricity and magnetism. Einstein resolved these problems by invoking a method of synchronizing clocks using the constant, finite speed of light as the maximum signal velocity. This led directly to the result that observers in motion relative to one another measure different elapsed times for the same event.

Spacetime

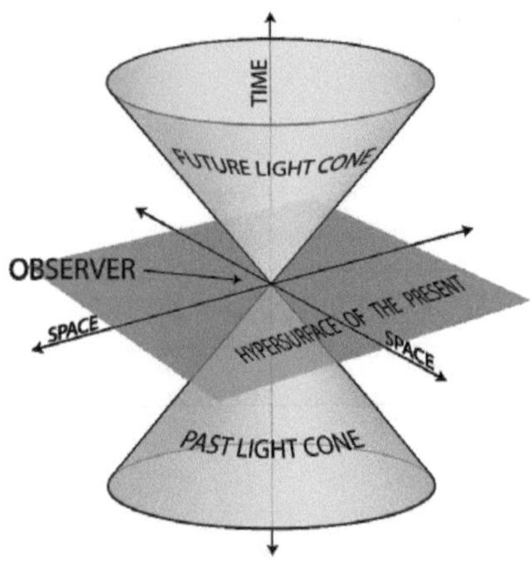

Two-dimensional space depicted in three-dimensional spacetime. The past and future light cones are absolute, the "present" is a relative concept different for observers in relative motion.

Time has historically been closely related with space, the two together comprising spacetime in Einstein's special relativity and general relativity. According to these theories, the concept of time depends on the spatial reference frame of the observer, and the human perception as well as the measurement by instruments such as clocks are different for observers in relative motion. In other words, if someone is flying in something that moves at the speed of light with a clock they do not notice a shift in time and see themselves as stationary because everything traveling at that speed slows down at the same rate (including the clock, the person's thought process, and the functions performed by their body). If there is an observer watching the other one travel, the person who is traveling at light speed begins to distort. However, for the person traveling, everything outside is shrunken down instead because they go around the earth's surface quickly and since everything is slower traveling can be done in shorter amount of time without the person noticing their timeframe has shifted. What seems like seconds for them can be hundreds of years for a person living in real time. Both times are correct for each person because the time is dependent on how fast you are traveling. However, the definition of each tense remains the same, the past is the set of events that can send light signals to the observer and the future is the set of events to which the observer can send light signals.

[58] [59] [60]

Time dilation

Einstein showed in his thought experiments that people travelling at different speeds, while agreeing on cause and effect, measures different time separations between events, and can even observe different chronological orderings between non-causally related events. Though these effects are typically minute in the human experience, the effect becomes much more pronounced for objects moving at speeds approaching the speed of light. Many subatomic particles exist for only a fixed fraction of a second in a lab relatively at rest, but some that travel close to the speed of light can be measured to travel farther and survive much longer than expected (a muon is one example). According to the special theory of relativity, in the high-speed particle's frame of reference, it exists, on the average, for a standard amount of time known as its mean lifetime, and the distance it travels in that time is zero, because its velocity is zero. Relative to a frame of reference at rest, time seems to "slow down" for the particle. Relative to the high-speed particle, distances seem to shorten. Einstein showed how both temporal and spatial dimensions can be altered (or "warped") by high-speed motion.

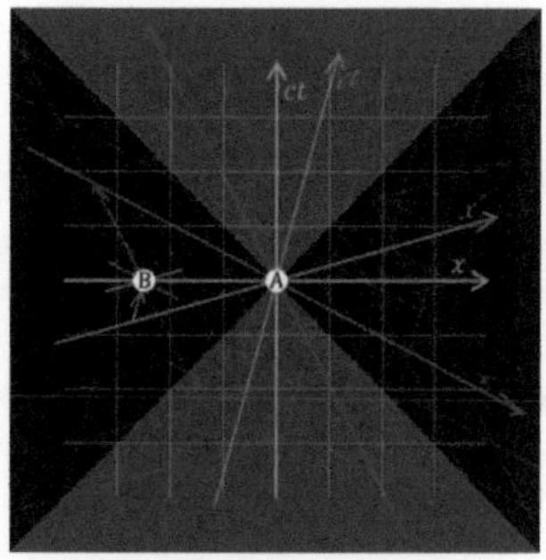

Relativity of simultaneity: Event B is simultaneous with A in the green reference frame, but it occurred before in the blue frame, and occurs later in the red frame.

Einstein (*The Meaning of Relativity*): "Two events taking place at the points A and B of a system K are simultaneous if they appear at the same instant when observed from the middle point, M, of the interval AB. Time is then defined as the ensemble of the indications of similar clocks, at rest relatively to K, which register the same simultaneously."

Einstein wrote in his book, *Relativity*, that simultaneity is also relative, i.e., two events that appear simultaneous to an observer in a particular inertial reference frame need not be judged as simultaneous by a second observer in a different inertial frame of reference.

Relativistic time versus Newtonian time

The animations visualise the different treatments of time in the Newtonian and the relativistic descriptions. At the heart of these differences are the Galilean and Lorentz transformations applicable in the Newtonian and relativistic theories, respectively.

In the figures, the vertical direction indicates time. The horizontal direction indicates distance (only one spatial dimension is taken into account), and the thick dashed curve is the spacetime trajectory ("world line") of the observer. The small dots indicate specific (past and future) events in spacetime.

The slope of the world line (deviation from being vertical) gives the relative velocity to the observer. Note how in both pictures the view of spacetime changes when the observer accelerates.

In the Newtonian description these changes are such that *time* is absolute: the movements of the observer do not influence whether an event occurs in the 'now' (i.e., whether an event passes the horizontal line through the observer).

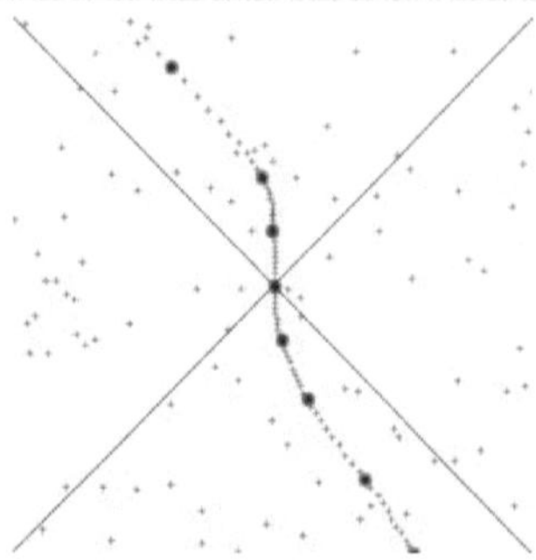
Views of spacetime along the world line of a rapidly accelerating observer in a relativistic universe. The events ("dots") that pass the two diagonal lines in the bottom half of the image (the past light cone of the observer in the origin) are the events visible to the observer.

However, in the relativistic description the *observability of events* is absolute: the movements of the observer do not influence whether an event passes the "light cone" of the observer. Notice that with the change from a Newtonian to a relativistic description, the concept of *absolute time* is no longer applicable: events move up-and-down in the figure depending on the acceleration of the observer.

Arrow of time

Time appears to have a direction – the past lies behind, fixed and immutable, while the future lies ahead and is not necessarily fixed. Yet for the most part the laws of physics do not specify an arrow of time, and allow any process to proceed both forward and in reverse. This is generally a consequence of time being modeled by a parameter in the system being analyzed, where there is no "proper time": the direction of the arrow of time is sometimes arbitrary. Examples of this include the Second law of thermodynamics, which states that entropy must increase over time (see Entropy); the cosmological arrow of time, which points away from the Big Bang, CPT symmetry, and the radiative arrow of time, caused by light only traveling forwards in time (see light cone). In particle physics, the violation of CP symmetry implies that there should be a small counterbalancing time asymmetry to preserve CPT symmetry as stated above. The standard description of measurement in quantum mechanics is also time asymmetric (see Measurement in quantum mechanics).

Quantised time

Time quantization is a hypothetical concept. In the modern established physical theories (the Standard Model of Particles and Interactions and General Relativity) time is not quantized.

Planck time (~ 5.4×10^{-44} seconds) is the unit of time in the system of natural units known as Planck units. Current established physical theories are believed to fail at this time scale, and many physicists expect that the Planck time might be the smallest unit of time that could ever be measured, even in principle. Tentative physical theories that describe this time scale exist; see for instance loop quantum gravity.

Time and the Big Bang

Stephen Hawking in particular has addressed a connection between time and the Big Bang. In *A Brief History of Time* and elsewhere, Hawking says that even if time did not begin with the Big Bang and there were another time frame before the Big Bang, no information from events then would be accessible to us, and nothing that happened then would have any effect upon the present time-frame.[61] Upon occasion, Hawking has stated that time actually began with the Big Bang, and that questions about what happened *before* the Big Bang are *meaningless*.[61] [62] [63] This less-nuanced, but commonly repeated formulation has received criticisms from philosophers such as Aristotelian philosopher Mortimer J. Adler.[64] [65]

Scientists have come to some agreement on descriptions of events that happened 10^{-35} seconds after the Big Bang, but generally agree that descriptions about what happened before one Planck time (5×10^{-44} seconds) after the Big Bang are likely to remain pure speculation.

Speculative physics beyond the Big Bang

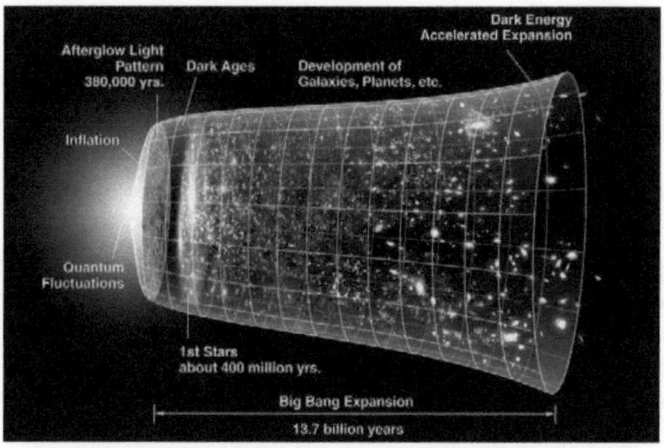

A graphical representation of the expansion of the universe with the inflationary epoch represented as the dramatic expansion of the metric seen on the left.

While the Big Bang model is well established in cosmology, it is likely to be refined in the future. Little is known about the earliest moments of the universe's history. The Penrose-Hawking singularity theorems require the existence of a singularity at the beginning of cosmic time. However, these theorems assume that general relativity is correct, but general relativity must break down before the universe reaches the Planck temperature, and a correct treatment of quantum gravity may avoid the singularity.[66]

There may also be parts of the universe well beyond what can be observed in principle. If inflation occurred this is likely, for exponential expansion would push large regions of space beyond our observable horizon.

Some proposals, each of which entails untested hypotheses, are:

- models including the Hartle–Hawking boundary condition in which the whole of space-time is finite; the Big Bang does represent the limit of time, but without the need for a singularity.[67]
- brane cosmology models[68] in which inflation is due to the movement of branes in string theory; the pre-big bang model; the ekpyrotic model, in which the Big Bang is the result of a collision between branes; and the cyclic model, a variant of the ekpyrotic model in which collisions occur periodically.[69] [70] [71]
- chaotic inflation, in which inflation events start here and there in a random quantum-gravity foam, each leading to a *bubble universe* expanding from its own big bang.[72]

Proposals in the last two categories see the Big Bang as an event in a much larger and older universe, or multiverse, and not the literal beginning.

Time travel

Time travel is the concept of moving backwards and/or forwards to different points in time, in a manner analogous to moving through space, and different from the normal "flow" of time to an earthbound observer. In this view, all points in time (including future times) "persist" in some way. Time travel has been a plot device in fiction since the 19th century. Traveling backwards in time has never been verified, presents many theoretic problems, and may be an impossibility.[73] Any technological device, whether fictional or hypothetical, that is used to achieve time travel is known as a time machine.

A central problem with time travel to the past is the violation of causality; should an effect precede its cause, it would give rise to the possibility of a temporal paradox. Some interpretations of time travel resolve this by accepting the possibility of travel between branch points, parallel realities, or universes.

Another solution to the problem of causality-based temporal paradoxes is that such paradoxes cannot arise simply because they have not arisen. As illustrated in numerous works of fiction, free will either ceases to exist in the past or the outcomes of such decisions are predetermined. As such, it would not be possible to enact the grandfather paradox because it is a historical fact that your grandfather was not killed before his child (your parent) was conceived. This view doesn't simply hold that history is an unchangeable constant, but that any change made by a hypothetical future time traveler would already have happened in his or her past, resulting in the reality that the traveler moves from.

More elaboration on this view can be found in the Novikov self-consistency principle.

Judgment of time

The specious present refers to the time duration wherein one's perceptions are considered to be in the present. The experienced present is said to be 'specious' in that, unlike the objective present, it is an interval and not a durationless instant. The term *specious present* was first introduced by the psychologist E.R. Clay, and later developed by William James.[74]

Biopsychology

The brain's judgement of time is known to be a highly distributed system, including at least the cerebral cortex, cerebellum and basal ganglia as its components. One particular component, the suprachiasmatic nuclei, is responsible for the circadian (or daily) rhythm, while other cell clusters appear capable of shorter-range (ultradian) timekeeping.

Psychoactive drugs can impair the judgement of time. Stimulants can lead both humans and rats to overestimate time intervals,[75] [76] while depressants can have the opposite effect.[77] The level of activity in the brain of neurotransmitters such as dopamine and norepinephrine may be the reason for this.[78] Such chemicals will either excite or inhibit the firing of neurons in the brain, with a greater firing rate allowing the brain to register the occurrence of more events within a given interval (speed up time) and a decreased firing rate reducing the brain's capacity to distinguish events occurring within a given interval (slow down time).[79]

Mental chronometry is the use of response time in perceptual-motor tasks to infer the content, duration, and temporal sequencing of cognitive operations.

Alterations

In addition to psychoactive drugs, judgements of time can be altered by temporal illusions (like the kappa effect[80]), age,[81] and hypnosis.[82] The sense of time is impaired in some people with neurological diseases such as Parkinson's disease and attention deficit disorder.

Psychologists assert that time seems to go faster with age, but the literature on this age-related perception of time remains controversial.[83] Those who support this notion point to the increased pressence of exciatory neurotransmitters in young people[79] and the role having new experiences plays in time perception.

Use of time

In sociology and anthropology, time discipline is the general name given to social and economic rules, conventions, customs, and expectations governing the measurement of time, the social currency and awareness of time measurements, and people's expectations concerning the observance of these customs by others. Arlie Russell Hochschild and Norbert Elias have written on the use of time from a sociological perspective.

The use of time is an important issue in understanding human behaviour, education, and travel behaviour. Time use research is a developing field of study. The question concerns how time is allocated across a number of activities (such as time spent at home, at work, shopping, etc.). Time use changes with technology, as the television or the Internet created new opportunities to use time in different ways. However, some aspects of time use are relatively stable over long periods of time, such as the amount of time spent traveling to work, which despite major changes in transport, has been observed to be about 20–30 minutes one-way for a large number of cities over a long period.

Time management is the organization of tasks or events by first estimating how much time a task requires and when it must be completed, and adjusting events that would interfere with its completion so it is done in the appropriate amount of time. Calendars and day planners are common examples of time management tools.

A sequence of events, or series of events, is a sequence of items, facts, events, actions, changes, or procedural steps, arranged in time order (chronological order), often with causality relationships among the items.[84] [85] [86] Because of causality, cause precedes effect, or cause and effect may appear together in a single item, but effect never precedes cause. A sequence of events can be presented in text, tables, charts, or timelines. The description of the items or events may include a timestamp. A sequence of events that includes the time along with place or location information to describe a sequential path may be referred to as a world line.

Uses of a sequence of events include stories,[87] historical events (chronology), directions and steps in procedures,[88] and timetables for scheduling activities. A sequence of events may also be used to help describe processes in science, technology, and medicine. A sequence of events may be focused on past events (e.g., stories, history, chronology), on future events that must be in a predetermined order (e.g., plans, schedules, procedures, timetables), or focused on the observation of past events with the expectation that the events will occur in the future (e.g., processes). The use of a sequence of events occurs in fields as diverse as machines (cam timer), documentaries (*Seconds From Disaster*), law (choice of law), computer simulation (discrete event simulation), and electric power transmission[89] (sequence of events recorder). A specific example of a sequence of events is the timeline of the Fukushima Daiichi nuclear disaster.

See also

- Horology
- Kairos
- Term (time)

Time's mortal aspect is personified in this bronze statue by Charles van der Stappen

Books

- *A Brief History of Time* by Stephen Hawking
- *About Time: Einstein's Unfinished Revolution* by Paul Davies
- *From Eternity to Here: The Quest for the Ultimate Theory of Time* by Sean M. Carroll
- *The Physical Basis of The Direction of Time* by Heinz-Dieter Zeh
- *An Experiment with Time* by J. W. Dunne
- *Einstein's Dreams* by Alan Lightman

Organizations

Leading scholarly organizations for researchers on the history and technology of time and timekeeping

- Antiquarian Horological Society – AHS (United Kingdom)
- Chronometrophilia (Switzerland)
- Deutsche Gesellschaft für Chronometrie – DGC (Germany)
- National Association of Watch and Clock Collectors - NAWCC (United States of America)

Miscellaneous arts and sciences	Miscellaneous units of time
• Anachronistic • Date and time notation by country • List of cycles • Network Time Protocol (NTP) • Nonlinear (arts) • Philosophy of physics • Rate (mathematics)	• Fiscal year • Half-life • Hexadecimal time • Season • Tithi • Unit of time • Unix epoch

References

[1] "Newton did for time what the Greek geometers did for space, idealized it into an exactly measurable dimension." *About Time: Einstein's Unfinished Revolution*, Paul Davies, p. 31, Simon & Schuster, 1996, ISBN 978-0684818221

[2] "Oxford Dictionaries:Time" (http://oxforddictionaries.com/definition/time). Oxford University Press. 2011. . Retrieved 2011-12-18. "the indefinite continued progress of existence and events in the past, present, and future regarded as a whole"

- "Webster's New World College Dictionary" (http://www.yourdictionary.com/time). 2010. . Retrieved 2011-04-09. "1.indefinite, unlimited duration in which things are considered as happening in the past, present, or future; every moment there has ever been or ever will be... a system of measuring duration 2.the period between two events or during which something exists, happens, or acts; measured or measurable interval"
- "The American Heritage Stedman's Medical Dictionary @dictionary.com" (http://dictionary.reference.com/browse/time?r=66). 2002. . Retrieved 2011-04-09. "A duration or relation of events expressed in terms of past, present, and future, and measured in units such as minutes, hours, days, months, or years."
- "Collins Language.com" (http://www.collinslanguage.com/results.aspx?context=3&reversed=False&action=define&homonym=-1&text=time). HarperCollins. 2011. . Retrieved 2011-12-18. "1. The continuous passage of existence in which events pass from a state of potentiality in the future, through the present, to a state of finality in the past. 2. *physics* a quantity measuring duration, usually with reference to a periodic process such as the rotation of the earth or the vibration of electromagnetic radiation emitted from certain atoms. In classical mechanics, time is absolute in the sense that the time of an event is independent of the observer. According to the theory of relativity it depends on the observer's frame of reference. Time is considered as a fourth coordinate required, along with three spatial coordinates, to specify an event."
- . 2002. . Retrieved 2011-04-09. "1. A continuous, measurable quantity in which events occur in a sequence proceeding from the past through the present to the future. 2a. An interval separating two points of this quantity; a duration. 2b. A system or reference frame in which such intervals are measured or such quantities are calculated."
- "Eric Weisstein's World of Science" (http://scienceworld.wolfram.com/physics/Time.html). 2007. . Retrieved 2011-04-09. "A quantity used to specify the order in which events occurred and measure the amount by which one event preceded or followed another. In special relativity, ct (where c is the speed of light and t is time), plays the role of a fourth dimension."

[4] "Time" (http://ahdictionary.com/word/search.html?q=time), *The American Heritage® Dictionary of the English Language* (Houghton Mifflin Company), 2011, , "A nonspatial continuum in which events occur in apparently irreversible succession from the past through the present to the future."

[5] Merriam-Webster Dictionary (http://www.merriam-webster.com/dictionary/time) the measured or measurable period during which an action, process, or condition exists or continues : duration; a nonspatial continuum which is measured in terms of events that succeed one another from past through present to future

[6] Compact Oxford English Dictionary A limited stretch or space of continued existence, as the interval between two successive events or acts, or the period through which an action, condition, or state continues. (1971)

- "Internet Encyclopedia of Philosophy" (http://www.iep.utm.edu/time/). 2010. . Retrieved 2011-04-09. "Time is what clocks measure. We use time to place events in sequence one after the other, and we use time to compare how long events last... Among philosophers of physics, the most popular short answer to the question "What is physical time?" is that it is not a substance or object but rather a special system of relations among instantaneous events. This working definition is offered by Adolf Grünbaum who applies the contemporary mathematical theory of continuity to physical processes, and he says time is a linear continuum of instants and is a distinguished one-dimensional sub-space of four-dimensional spacetime."
- "Dictionary.com Unabridged, based on Random House Dictionary" (http://dictionary.reference.com/browse/time?r=66). 2010. . Retrieved 2011-04-09. "1. the system of those sequential relations that any event has to any other, as past, present, or future; indefinite and continuous duration regarded as that in which events succeed one another.... 3. (sometimes initial capital letter) a system or method of measuring or reckoning the passage of time: mean time; apparent time; Greenwich Time. 4. a limited period or interval, as between two successive events: a long time.... 14. a particular or definite point in time, as indicated by a clock: What time is it? ... 18. an indefinite, frequently prolonged period or duration in the future: Time will tell if what we have done here today was right."
- Ivey, Donald G.; Hume, J.N.P. (1974). *Physics* (http://books.google.com/books?id=HrHvAAAAMAAJ&q="time+is+what+clocks+measure"&dq="time+is+what+clocks+measure"&hl=en). **1**. Ronald Press. p. 65. . "Our operational definition of time is that time is what clocks measure."

[8] Le Poidevin, Robin (Winter 2004). "The Experience and Perception of Time" (http://plato.stanford.edu/archives/win2004/entries/time-experience). In Edward N. Zalta. *The Stanford Encyclopedia of Philosophy*. . Retrieved 2011-04-09.

[9] Sean M Carroll (2009). *From Eternity to Here: The Quest for the Ultimate Theory of Time* (http://books.google.com/books?id=Uak1wtcXrjwC&redir_esc=y). Dutton. ISBN 978-0-525-95133-9. .

[10] Adam Frank, *Cosmology and Culture at the Twilight of the Big Bang*, "the time we imagined from the cosmos and the time we imagined into the human experience turn out to be woven so tightly together that we have lost the ability to see each of them for what it is." p. xv, Free Press, 2011, ISBN 978-1439169599

[11] St. Augustine, *Confessions*, Simon & Brown, 2012, ISBN 978-1613823262

[12] Official Baseball Rules, 2011 Edition (2011). "Rules 8.03 and 8.04" (http://mlb.mlb.com/mlb/downloads/y2011/Official_Baseball_Rules.pdf) (Free PDF download). Major League Baseball. . Retrieved 2012-07-07. "Rule 8.03 Such preparatory pitches shall not consume more than one minute of time...Rule 8.04 When the bases are unoccupied, the pitcher shall deliver the ball to the batter within 12 seconds...The 12-second timing starts when the pitcher is in possession of the ball and the batter is in the box, alert to the pitcher. The timing stops when the pitcher releases the ball"

[13] "Guinness Book of Baseball World Records" (http://www.baseball-almanac.com/recbooks/rb_guin.shtml) (web). Guinness World Records, Ltd.. . Retrieved 2012-07-07. "The record for the fastest time for circling the bases is 13.3 seconds, set by Evar Swanson at Columbus, Ohio in 1932...The greatest reliably recorded speed at which a baseball has been pitched is 100.9 mph by Lynn Nolan Ryan (California Angels) at Anaheim Stadium in California on August 20, 1974."

[14] Zeigler, Kenneth (2008). *Getting organized at work : 24 lessons to set goals, establish priorities, and manage your time* (http://books.google.com/books?id=acPPD6lCCxcC&printsec=frontcover&dq=Getting+organized+at+work+:+24+lessons+to+set+goals,+establish+priorities,+and+manage+your+time&hl=en&sa=X&ei=2zH5T52FLfCf6QGHnrT9Bg&ved=0CDYQuwUwAA#v=onepage&q=Getting organized at work : 24 lessons to set goals, establish priorities, and manage your time&f=false). McGraw-Hill. ISBN 9780071591386. . 108 pages

[15] Burnham, Douglas : Staffordshire University (2006). "Gottfried Wilhelm Leibniz (1646–1716) Metaphysics – 7. Space, Time, and Indiscernibles" (http://www.iep.utm.edu/leib-met/#H7). *The Internet Encyclopedia of Philosophy*. . Retrieved 2011-04-09. "First of all, Leibniz finds the idea that space and time might be substances or substance-like absurd (see, for example, "Correspondence with Clarke," Leibniz's Fourth Paper, §8ff). In short, an empty space would be a substance with no properties; it will be a substance that even God cannot modify or destroy.... That is, space and time are internal or intrinsic features of the complete concepts of things, not extrinsic.... Leibniz's view has two major implications. First, there is no absolute location in either space or time; location is always the situation of an object or event relative to other objects and events. Second, space and time are not in themselves real (that is, not substances). Space and time are, rather, ideal. Space and time are just metaphysically illegitimate ways of perceiving certain virtual relations between substances. They are phenomena or, strictly speaking, illusions (although they are illusions that are well-founded upon the internal properties of substances).... It is sometimes convenient to think of space and time as something "out there," over and above the entities and their relations to each other, but this convenience must not be confused with reality. Space is nothing but the order of co-existent objects; time nothing but the order of successive events. This is usually called a relational theory of space and time."

[16] Cummings, Raymond King (1922). *The Girl in the Golden Atom* (http://books.google.com/?id=tA647bGiWwsC&pg=PA46&lpg=PA46&dq="keeps+everything"#v=onepage&q="keeps everything"). U of Nebraska Press. p. 46. ISBN 978-0-8032-6457-1. . Retrieved 2011-04-09. Chapter 5. Cummings repeated this sentence in several of his novellas. Sources, such as this one (http://books.google.com/books?id=39KQY1FnSfkC&pg=PA98&lpg=PA98&dq=cummings+"time+professor"+"keeps+everything"#v=onepage&q=cummings "time professor" "keeps everything"&f=false), attribute it to his earlier work, *The Time Professor*, in 1921. Before taking book form, several of Cummings's stories appeared serialized in magazines. The first eight chapters of his *The Girl in the Golden Atom* appeared (http://www.isfdb.org/cgi-bin/title.cgi?46112) in *All-Story Magazine* on 15 March 1919. In the novel version (http://www.gutenberg.org/ebooks/21094) the quote about time appears in Chapter V (http://www.gutenberg.org/files/21094/21094-h/21094-h.htm#CHAPTER_V).

[17] International, Rotary (Aug 1973). *The Rotarian* (http://books.google.com/?id=vjUEAAAAMBAJ&lpg=PP1). Published by Rotary International. p. 47. ISSN 0035-838X. . Retrieved 2011-04-09., What does a man possess? page 47 (http://books.google.com/books?id=vjUEAAAAMBAJ&lpg=PP1&pg=PA47)

[18] Daintith, John (2008). *Biographical Encyclopedia of Scientists* (http://books.google.com/?id=vqTNfnKJVPAC) (third ed.). CRC Press. p. 796. ISBN 1-4200-7271-4. . Retrieved 2011-04-09., Page 796, quoting Wheeler from the American Journal of Physics, 1978 (http://books.google.com/books?id=vqTNfnKJVPAC&pg=PA796)

[19] Davies, Davies (1995). *About time: Einstein's unfinished revolution* (http://books.google.com/?id=SZPuAAAAMAAJ). Simon & Schuster. p. 236. ISBN 0-671-79964-9. . Retrieved 2011-04-09.

[20] Rynasiewicz, Robert : Johns Hopkins University (12 August 2004). "Newton's Views on Space, Time, and Motion" (http://plato.stanford.edu/entries/newton-stm/). *Stanford Encyclopedia of Philosophy*. Stanford University. . Retrieved 2012-02-05. "Newton did not regard space and time as genuine substances (as are, paradigmatically, bodies and minds), but rather as real entities with their own manner of existence as necessitated by God's existence... To paraphrase: Absolute, true, and mathematical time, from its own nature, passes equably without relation to anything external, and thus without reference to any change or way of measuring of time (e.g., the hour, day, month, or year)."

[21] Markosian, Ned. "Time" (http://plato.stanford.edu/entries/time/#3). In Edward N. Zalta. *The Stanford Encyclopedia of Philosophy (Winter 2002 Edition)*. . Retrieved 2011-09-23. "The opposing view, normally referred to either as "Platonism with Respect to Time" or as "Absolutism with Respect to Time," has been defended by Plato, Newton, and others. On this view, time is like an empty container into which events may be placed; but it is a container that exists independently of whether or not anything is placed in it."

[22] Mattey, G. J. : UC Davis (22 January 1997). "Critique of Pure Reason, Lecture notes: Philosophy 175 UC Davis" (http://www-philosophy.ucdavis.edu/mattey/kant/TIMELEC.HTM). . Retrieved 2011-04-09. "What is correct in the Leibnizian view was its anti-metaphysical stance. Space and time do not exist in and of themselves, but in some sense are the product of the way we represent things. The[y] are ideal, though not in the sense in which Leibniz thought they are ideal (figments of the imagination). The ideality of space is its mind-dependence: it is only a condition of sensibility.... Kant concluded "absolute space is not an object of outer sensation; it is rather a fundamental concept which first of all makes possible all such outer sensation."...Much of the argumentation pertaining to space is applicable, mutatis mutandis, to time, so I will not rehearse the arguments. As space is the form of outer intuition, so time is the form of inner intuition.... Kant claimed that time is real, it is "the real form of inner intuition.""

[23] McCormick, Matt : California State University, Sacramento (2006). "Immanuel Kant (1724–1804) Metaphysics: 4. Kant's Transcendental Idealism" (http://www.iep.utm.edu/kantmeta/#H4). *The Internet Encyclopedia of Philosophy*. . Retrieved 2011-04-09. "Time, Kant argues, is also necessary as a form or condition of our intuitions of objects. The idea of time itself cannot be gathered from experience because succession and simultaneity of objects, the phenomena that would indicate the passage of time, would be impossible to represent if we did not already possess the capacity to represent objects in time.... Another way to put the point is to say that the fact that the mind of the knower makes the *a priori* contribution does not mean that space and time or the categories are mere figments of the imagination. Kant is an empirical realist about the world we experience; we can know objects as they appear to us. He gives a robust defense of science and the study of the natural world from his argument about the mind's role in making nature. All discursive, rational beings must conceive of the physical world as spatially and temporally unified, he argues."

[24] Duff, Okun, Veneziano, *ibid.* p. 3. "There is no well established terminology for the fundamental constants of Nature. ... The absence of accurately defined terms or the uses (i.e., actually misuses) of ill-defined terms lead to confusion and proliferation of wrong statements."

[25] Carrol, Sean, Chapter One, Section Two, Plume, 2010. "From Eternity to Here". ISBN 978-0452296541. "As human beings we 'feel' the passage of time."

[26] Lehar, Steve. (2000). The Function of Conscious Experience: An Analogical Paradigm of Perception and Behavior (http://cns-alumni.bu.edu/~slehar/webstuff/consc1/consc1.html), *Consciousness and Cognition*.

[27] Richards, E. G. (1998). *Mapping Time: The Calendar and its History*. Oxford University Press. pp. 3–5.

[28] Rudgley, Richard (1999). *The Lost Civilizations of the Stone Age*. New York: Simon & Schuster. pp. 86–105.

[29] Barnett, Jo Ellen *Time's Pendulum: The Quest to Capture Time—from Sundials to Atomic Clocks* Plenum, 1998 ISBN 0-306-45787-3 p.28

[30] Barnett, *ibid*, p.37

[31] Laurence Bergreen, *Over the Edge of the World: Magellan's Terrifying Circumnavigation of the Globe*, HarperCollins Publishers, 2003, hardcover 480 pages, ISBN 0-06-621173-5

[32] North, J. (2004) *God's Clockmaker: Richard of Wallingford and the Invention of Time*. Oxbow Books. ISBN 1-85285-451-0

[33] Watson, E (1979) "The St Albans Clock of Richard of Wallingford". *Antiquarian Horology* 372–384.

[34] "NIST Unveils Chip-Scale Atomic Clock" (http://www.nist.gov/public_affairs/releases/miniclock.cfm). 27 August 2004. . Retrieved 2011-06-09.

[35] "New atomic clock can keep time for 200 million years: Super-precise instruments vital to deep space navigation" (http://www.canada.com/vancouversun/news/story.html?id=e24ccfa7-44eb-40b7-8b67-daf8263569ff). *Vancouver Sun*. 16 February 2008. . Retrieved 2011-04-09.

[36] "Byrhtferth of Ramsey". (2008). In Encyclopædia Britannica. Retrieved 2008-09-15, from Encyclopædia Britannica Online: http://search.eb.com/eb/article-9438957

[37] "atom", Oxford English Dictionary, Draft Revision Sept. 2008 (contains relevant citations from Byrhtferth's *Enchiridion*)

[38] "12 attoseconds is the world record for shortest controllable time" (http://www.physorg.com/news192909576.html). 12 May 2010. . Retrieved 2012-04-19.

[39] McCarthy, Dennis D.; Seidelmann, P. Kenneth (2009). *Time: from Earth rotation to atomic physics* (http://books.google.com/books?id=7NdrK4e77CIC). Wiley-VCH. p. 18. ISBN 3-527-40780-4. ., Extract of page 18 (http://books.google.com/books?id=7NdrK4e77CIC&pg=PA18)

[40] Jones, Floyd Nolen (2005). *The Chronology of the Old Testament* (http://books.google.com/books?id=ZkBasQYRy4sC) (15th ed.). New Leaf Publishing Group. p. 287. ISBN 0-89051-416-X. ., Extract of page 287 (http://books.google.com/books?id=ZkBasQYRy4sC&pg=PA287)

[41] Organisation Intergouvernementale de la Convention du Métre (1998) (PDF). *The International System of Units (SI), 7th Edition* (http://www1.bipm.org/utils/en/pdf/si-brochure.pdf). . Retrieved 2011-04-09.

[42] "Base unit definitions: Second" (http://physics.nist.gov/cuu/Units/second.html). NIST. . Retrieved 2011-04-09.

[43] Navratil, Gerhard (2009). *Research Trends in Geographic Information Science* (http://books.google.be/books?id=q8w728aa1CkC). Springer Japan. p. 217. ISBN 3-540-88243-X. . Retrieved 2011-04-09.

[44] Layton, Robert (1994). *Who needs the past?: indigenous values and archaeology* (http://books.google.com/books?id=7TPIDL9RdsoC) (2nd ed.). Routledge. p. 7. ISBN 0-415-09558-1. . Retrieved 2011-04-09., Introduction, p. 7 (http://books.google.com/books?id=7TPIDL9RdsoC&pg=PA7)

[45] Dagobert Runes, *Dictionary of Philosophy*, p. 318

[46] Augustine of Hippo. *Confessions* (http://en.wikisource.org/wiki/Nicene_and_Post-Nicene_Fathers:_Series_I/Volume_I/Confessions/Book_XI/Chapter_14). . Retrieved 2011-04-09. Book 11, Chapter 14.

[47] Gottfried Martin, *Kant's Metaphysics and Theory of Science*

[48] Kant, Immanuel (1787). *The Critique of Pure Reason, 2nd edition* (http://ebooks.adelaide.edu.au/k/kant/immanuel/k16p/). . Retrieved 2011-04-09. translated by J. M. D. Meiklejohn, eBooks@Adelaide, 2004

[49] Bergson, Henri (1907) *Creative Evolution*. trans. by Arthur Mitchell. Mineola: Dover, 1998.

[50] Balslev, Anindita N.; and Jitendranath Mohanty (November 1992). *Religion and Time* (http://books.google.com/?id=y94cKeEVa3sC&printsec=frontcover&dq=Religion+and+time#v=onepage&q=Heidegger&f=false). Studies in the History of Religions, 54.. The Netherlands: Brill Academic Publishers. pp. 53, 54, 55, 56, 57, 58, and 59. ISBN 978-90-04-09583-0. .

[51] Martin Heidegger (1962). "V" (http://books.google.com/?id=S57m5gW0L-MC&pg=PA425&lpg=PA425&dq=heidegger+sequence#v=onepage&q=heidegger sequence&f=false). *Being and Time*. p. 425. ISBN 978-0-631-19770-6. .

[52] Harry Foundalis. "You are about to disappear" (http://www.foundalis.com/phi/WhyTimeFlows.htm). . Retrieved 2011-04-09.

[53] Huston, Tom. "Buddhism and the illusion of time" (http://www.buddhasvillage.com/teachings/time.htm). . Retrieved 2011-04-09.

[54] Garfield, Jay L. (1995). *The fundamental wisdom of the middle way: Nāgārjuna's Mūlamadhyamakakārikā* (http://books.google.com/?id=kfsyfoO1lIYC&pg=RA1-PR19&dq=The+fundamental+wisdom+of+the+middle+way+time#v=onepage&q&f=false). New York: Oxford University Press. ISBN 978-0-19-509336-0. .

[55] "Time is an illusion?" (http://physicsandphysicists.blogspot.com/2007/03/time-is-illusion.html). . Retrieved 2011-04-09.

[56] Herman M. Schwartz, *Introduction to Special Relativity*, McGraw-Hill Book Company, 1968, hardcover 442 pages, see ISBN 0-88275-478-5 (1977 edition), pp. 10–13

[57] A. Einstein, H. A. Lorentz, H. Weyl, H. Minkowski, *The Principle of Relativity*, Dover Publications, Inc, 2000, softcover 216 pages, ISBN 0-486-60081-5, See pp. 37–65 for an English translation of Einstein's original 1905 paper.

[58] http://www.youtube.com/watch?v=ev9zrt__lec

[59] http://www.youtube.com/watch?v=V7vpw4AH8QQ

[60] http://www.cracked.com/article_19659_7-theories-time-that-would-make-doc-browns-head-explode.html

[61] Hawking, Stephen (1996). "The Beginning of Time" (http://www.hawking.org.uk/the-beginning-of-time.html). University of Cambridge. . Retrieved 2012-07-08. "Since events before the Big Bang have no observational consequences, one may as well cut them out of the theory, and say that time began at the Big Bang. Events before the Big Bang, are simply not defined, because there's no way one could measure what happened at them. This kind of beginning to the universe, and of time itself, is very different to the beginnings that had been considered earlier."

[62] Hawking, Stephen (27 February 2006). "Professor Stephen Hawking lectures on the origin of the universe" (http://www.ox.ac.uk/media/news_stories/2006/060227.html). University of Oxford. . Retrieved 2012-12-05. "Suppose the beginning of the universe was like the South Pole of the earth, with degrees of latitude playing the role of time. The universe would start as a point at the South Pole. As one moves north, the circles of constant latitude, representing the size of the universe, would expand. To ask what happened before the beginning of the universe would become a meaningless question because there is nothing south of the South Pole."

[63] Ghandchi, Sam : Editor/Publisher (16 January 2004). "Space and New Thinking" (http://www.ghandchi.com/312-SpaceEng.htm). . Retrieved 2011-04-09. "and as Stephen Hawking puts it, asking what was before Big Bang is like asking what is North of North Pole, a meaningless question."

[64] Adler, Mortimer J., PhD. "Natural Theology, Chance, and God" (http://radicalacademy.com/adlertheology1.htm). . Retrieved 2011-04-09. "Hawking could have avoided the error of supposing that time had a beginning with the Big Bang if he had distinguished time as it is measured by physicists from time that is not measurable by physicists.... an error shared by many other great physicists in the twentieth century, the error of saying that what cannot be measured by physicists does not exist in reality."

"The Great Ideas Today". *Encyclopædia Britannica*. 1992.

[65] Adler, Mortimer J., PhD. "Natural Theology, Chance, and God" (http://radicalacademy.com/adlertheology2.htm). . Retrieved 2011-04-09. "Where Einstein had said that what is not measurable by physicists is of no interest to them, Hawking flatly asserts that what is not measurable by physicists does not exist—has no reality whatsoever.
With respect to time, that amounts to the denial of psychological time which is not measurable by physicists, and also to everlasting time—time before the Big Bang—which physics cannot measure. Hawking does not know that both Aquinas and Kant had shown that we cannot rationally establish that time is either finite or infinite."

"The Great Ideas Today". *Encyclopædia Britannica*. 1992.

[66] Hawking, Stephen; and Ellis, G. F. R. (1973). *The Large Scale Structure of Space-Time*. Cambridge: Cambridge University Press. ISBN 0-521-09906-4.

[67] J. Hartle and S. W. Hawking (1983). "Wave function of the universe". *Phys. Rev. D* **28** (12): 2960. Bibcode 1983PhRvD..28.2960H. doi:10.1103/PhysRevD.28.2960.

[68] Langlois, David (2002). "Brane cosmology: an introduction". *Progress of Theoretical Physics Supplement* **148**: 181. arXiv:hep-th/0209261. Bibcode 2002PThPS.148..181L. doi:10.1143/PTPS.148.181.

[69] Linde, Andre (2002). "Inflationary Theory versus Ekpyrotic/Cyclic Scenario". *In: the future of theoretical physics and cosmology. Edited by G. W. Gibbons*: 801. arXiv:hep-th/0205259. Bibcode 2003ftpc.book..801L.

[70] "Recycled Universe: Theory Could Solve Cosmic Mystery" (http://www.space.com/2372-recycled-universe-theory-solve-cosmic-mystery.html). Space.com. 8 May 2006. . Retrieved 2011-04-09.

[71] "What Happened Before the Big Bang?" (http://web.archive.org/web/20070704150957/http://www.science.psu.edu/alert/Bojowald6-2007.htm). Archived from the original (http://www.science.psu.edu/alert/Bojowald6-2007.htm) on 2007-07-04. . Retrieved 2011-04-09.

[72] A. Linde (1986). "Eternal chaotic inflation". *Mod. Phys. Lett.* **A1** (2): 81. Bibcode 1986MPLA....1...81L. doi:10.1142/S0217732386000129. A. Linde (1986). "Eternally existing self-reproducing chaotic inflationary universe". *Phys. Lett.* **B175** (4): 395–400. Bibcode 1986PhLB..175..395L. doi:10.1016/0370-2693(86)90611-8.

[73] G. Quznetsov, Prespacetime Journal, March 2010, Vol.1, Issue 2, Page 274-275 (http://prespacetime.com/index.php/pst/article/view/10)

[74] Andersen, Holly; Rick Grush (pending) (PDF). *A brief history of time-consciousness: historical precursors to James and Husserl* (http://mind.ucsd.edu/papers/bhtc/Andersen&Grush.pdf). Journal of the History of Philosophy. . Retrieved 2011-04-09.

[75] Wittmann, M.; Leland DS, Churan J, Paulus MP. (8 October 2007). "Impaired time perception and motor timing in stimulant-dependent subjects" (online abstract). *Drug Alcohol Depend.* **90** (2–3): 183–92. doi:10.1016/j.drugalcdep.2007.03.005. PMC 1997301. PMID 17434690.

[76] Cheng, Ruey-Kuang; Macdonald, Christopher J.; Meck, Warren H. (2006). "Differential effects of cocaine and ketamine on time estimation: Implications for neurobiological models of interval timing" (http://cat.inist.fr/?aModele=afficheN&cpsidt=18303059) (online abstract). *Pharmacology, biochemistry and behavior* **85** (1): 114–122. doi:10.1016/j.pbb.2006.07.019. PMID 16920182. . Retrieved 2011-04-09.

[77] Tinklenberg, Jared R.; Walton T. Roth1; Bert S. Kopell (January 1976). "Marijuana and ethanol: Differential effects on time perception, heart rate, and subjective response" (http://www.springerlink.com/content/q1227453r481x439/). *Psychopharmacology* **49** (3): 275–279. doi:10.1007/BF00426830. PMID 826945. . Retrieved 2011-04-09.

[78] Arzy, Shahar; Istvan Molnar-Szakacs; Olaf Blanke (18 June 2008). "Self in Time: Imagined Self-Location Influences Neural Activity Related to Mental Time Travel" (http://www.jneurosci.org/content/28/25/6502.abstract) (Abstract). *The Journal of Neuroscience* **28** (25): 6502–6507. doi:10.1523/JNEUROSCI.5712-07.2008. PMID 18562621. . Retrieved 2011-04-09.

[79] Carter, Rita. *The Human Brain Book*. pp. 186–187.

[80] Wada Y, Masuda T, Noguchi K, 2005, "Temporal illusion called 'kappa effect' in event perception" Perception 34 ECVP Abstract Supplement

[81] Robert, Adler. "Look how time flies.." (http://www.newscientist.com/article/mg16422180.900-look-how-time-flies). . Retrieved 2011-04-09.

[82] Bowers, Kenneth; Brenneman, HA (January 1979). "Hypnosis and the perception of time" (http://www.informaworld.com/smpp/content~content=a790232921~db=all). *International Journal of Clinical and Experimental Hypnosis* (International Journal of Clinical and Experimental Hypnosis) **27** (1): 29–41. doi:10.1080/00207147908407540. PMID 541126.

[83] Gruber, Ronald P.; Wagner, Lawrence F.; Block, Richard A. (2000). "Subjective Time Versus Proper (Clock) Time" (http://books.google.com/books?id=LMsDqsvcxckC). In Buccheri, R.; Di Gesù, V.; Saniga, Metod. *Studies on the structure of time: from physics to psycho(patho)logy*. Springer. p. 54. ISBN 0-306-46439-X. . Retrieved 2011-04-09., Extract of page 54 (http://books.google.com/books?id=LMsDqsvcxckC&pg=PA54)

[84] "Sequence - Order of Important Events" (http://www.austinschools.org/curriculum/la/resources/documents/instResources/LA_res_Seq_ORS_Module.pdf). Austin Independent School District. 2009. .

[85] "Sequence of Events Worksheets" (http://www.reference.com/motif/Science/sequence-of-events-worksheets). Reference.com. .

[86] Compiled by David Luckham and Roy Schulte. "Event Processing Glossary – Version 2.0" (http://www.complexevents.com/2011/08/23/event-processing-glossary-version-2-0/). Complex Event Processing. .

[87] Richard Nordquist. "narrative" (http://grammar.about.com/od/mo/g/narrative2term.htm). About.com. .

[88] David J. Piasecki. "Inventory Accuracy Glossary" (http://www.accuracybook.com/glossary.htm). AccuracyBook.com (OPS Publishing). .

[89] "Utility Communications Architecture (UCA) glossary" (http://www.nettedautomation.com/glossary_menue/glossary_uca.html). NettedAutomation. .

Further reading

- Barbour, Julian (1999). *The End of Time: The Next Revolution in Our Understanding of the Universe*. Oxford University Press. ISBN 0-19-514592-5.
- Landes, David (2000). *Revolution in Time*. Harvard University Press. ISBN 0-674-00282-2.
- Das, Tushar Kanti (1990). *The Time Dimension: An Interdisciplinary Guide*. New York: Praeger. ISBN 0-275-92681-8.- Research bibliography
- Davies, Paul (1996). *About Time: Einstein's Unfinished Revolution*. New York: Simon & Schuster Paperbacks. ISBN 0-684-81822-1.
- Feynman, Richard (1994) [1965]. *The Character of Physical Law* (http://mitpress.mit.edu/catalog/item/default.asp?ttype=2&tid=5277). Cambridge (Mass): The MIT Press. pp. 108–126. ISBN 0-262-56003-8.
- Galison, Peter (1992). *Einstein's Clocks and Poincaré's Maps: Empires of Time*. New York: W. W. Norton. ISBN 0-393-02001-0.
- Highfield, Roger (1992). *Arrow of Time: A Voyage through Science to Solve Time's Greatest Mystery*. Random House. ISBN 0-449-90723-6.
- Mermin, N. David (2005). *It's About Time: Understanding Einstein's Relativity* (http://press.princeton.edu/titles/8112.html). Princeton University Press. ISBN 0-691-12201-6.
- Penrose, Roger (1999) [1989]. *The Emperor's New Mind: Concerning Computers, Minds, and the Laws of Physics* (http://ukcatalogue.oup.com/product/9780192861986.do). New York: Oxford University Press. pp. 391–417. ISBN 0-19-286198-0. Retrieved 2011-04-09.

- Price, Huw (1996). *Time's Arrow and Archimedes' Point* (http://sydney.edu.au/time/price/TAAP.html). Oxford University Press. ISBN 0-19-511798-0. Retrieved 2011-04-09.
- Reichenbach, Hans (1999) [1956]. *The Direction of Time* (http://store.doverpublications.com/0486409260.html). New York: Dover. ISBN 0-486-40926-0.
- Stiegler, Bernard, *Technics and Time, 1: The Fault of Epimetheus*
- Quznetsov, Gunn A. (2006). *Logical Foundation of Theoretical Physics*. Nova Sci. Publ.. ISBN 1-59454-948-6.
- Whitrow, Gerald J. (1973). *The Nature of Time*. Holt, Rinehart and Wilson (New York).
- Whitrow, Gerald J. (1980). *The Natural Philosophy of Time*. Clarendon Press (Oxford).
- Whitrow, Gerald J. (1988). *Time in History. The evolution of our general awareness of time and temporal perspective*. Oxford University Press. ISBN 0-19-285211-6.
- Rovelli, Carlo (2006). *What is time? What is space?* (http://www.direnzo.it/main.phtml?Language=en&Doc=0001&ISBN=8883231465). Rome: Di Renzo Editore. ISBN 88-8323-146-5.
- Charlie Gere, (2005) *Art, Time and Technology: Histories of the Disappearing Body*, Berg
- Craig Callendar, *Introducing Time*, Icon Books, 2010, ISBN 978-1848311206
- Benjamin Gal-Or, *Cosmology, Physics and Philosophy*, Springer Verlag, 1981, 1983, 1987, ISBN 0-387-90581-2, ISBN 0-387-96526-2.

External links

- Exploring Time (http://exploringtime.org/?page=segments) from Planck Time to the lifespan of the universe
- Different systems of measuring time (http://tycho.usno.navy.mil/systime.html)
- Time (http://www.bbc.co.uk/programmes/p005465z) on *In Our Time* at the BBC. (listen now (http://www.bbc.co.uk/iplayer/console/p005465z/In_Our_Time_Time))
- Dowden, Bradley (California State University, Sacramento) (2007). "Time" (http://www.iep.utm.edu/time/). In James Fieser, PhD, Bradley Dowden, PhD. *The Internet Encyclopedia of Philosophy*. Retrieved 2011-04-09.
- Le Poidevin, Robin (Winter 2004). "The Experience and Perception of Time" (http://plato.stanford.edu/archives/win2004/entries/time-experience). In Edward N. Zalta. *The Stanford Encyclopedia of Philosophy*. Retrieved 2011-04-09.
- Time at Open Directory (http://www.dmoz.org/Reference/Time/)

Distributed_control_system

Part of a series of articles on
Industry

Manufacturing methods Batch production • Job production Flow production
Improvement methods LM • TPM • QRM • VDM TOC • Six Sigma • RCM
Information & communication ISA-88 • ISA-95 • ERP SAP • IEC 62264 • B2MML
Process control PLC • DCS

A **distributed control system** (DCS) refers to a control system usually of a manufacturing system, process or any kind of dynamic system, in which the controller elements are not central in location (like the brain) but are distributed throughout the system with each component sub-system controlled by one or more controllers.

DCS (Distributed Control System) is a computerized control system used to control the production line in the industry

The entire system of controllers is connected by networks for communication and monitoring.

DCS is a very broad term used in a variety of industries, to monitor and control distributed equipment.

- Electrical power grids and electrical generation plants
- Environmental control systems
- Traffic signals
- Radio signals
- Water management systems
- Oil refining plants
- Metallurgical process plants
- Chemical plants
- Pharmaceutical manufacturing
- Sensor networks
- Dry cargo and bulk oil carrier ships

Elements

A DCS typically uses custom designed processors as controllers and uses both proprietary interconnections and communications protocol for communication. Input and output modules form component parts of the DCS. The processor receives information from input modules and sends information to output modules. The input modules receive information from input instruments in the process (or field) and transmit instructions to the output instruments in the field. Computer buses or electrical buses connect the processor and modules through multiplexer or demultiplexers. Buses also connect the distributed controllers with the central controller and finally to the Human–machine interface (HMI) or control consoles. See Process automation system.

The elements of a DCS may connect directly to physical equipment such as switches, pumps and valves or they may work through an intermediate system such as a SCADA system.

Applications

Distributed control systems (DCSs) are dedicated systems used to control manufacturing processes that are continuous or batch-oriented, such as oil refining, petrochemicals, central station power generation, fertilizers, pharmaceuticals, food and beverage manufacturing, cement production, steelmaking, and papermaking. DCSs are connected to sensors and actuators and use setpoint control to control the flow of material through the plant. The most common example is a setpoint control loop consisting of a pressure sensor, controller, and control valve. Pressure or flow measurements are transmitted to the controller, usually through the aid of a signal conditioning input/output (I/O) device. When the measured variable reaches a certain point, the controller instructs a valve or actuation device to open or close until the fluidic flow process reaches the desired setpoint. Large oil refineries have many thousands of I/O points and employ very large DCSs. Processes are not limited to fluidic flow through pipes, however, and can also include things like paper machines and their associated quality controls (see quality control system QCS), variable speed drives and motor control centers, cement kilns, mining operations, ore processing facilities, and many others.

A typical DCS consists of functionally and/or geographically distributed digital controllers capable of executing from 1 to 256 or more regulatory control loops in one control box. The input/output devices (I/O) can be integral with the controller or located remotely via a field network. Today's controllers have extensive computational capabilities and, in addition to proportional, integral, and derivative (PID) control, can generally perform logic and sequential control. Modern DCSs also support neural networks and fuzzy application.

DCS systems are usually designed with redundant processors to enhance the reliability of the control system. Most systems come with canned displays and configuration software which enables the end user to set up the control system without a lot of low level programming. This allows the user to better focus on the application rather than the equipment, although a lot of system knowledge and skill is still required to support the hardware and software as well as the applications. Many plants have dedicated groups that focus on this task. These groups are in many cases augumented by vendor support personnel and/or maintenance support contracts.

DCSs may employ one or more workstations and can be configured at the workstation or by an off-line personal computer. Local communication is handled by a control network with transmission over twisted pair, coaxial, or fiber optic cable. A server and/or applications processor may be included in the system for extra computational, data collection, and reporting capability.

History

Early minicomputers were used in the control of industrial processes since the beginning of the 1960s. The IBM 1800, for example, was an early computer that had input/output hardware to gather process signals in a plant for conversion from field contact levels (for digital points) and analog signals to the digital domain.

The first industrial control computer system was built 1959 at the Texaco Port Arthur, Texas, refinery with an RW-300 of the Ramo-Wooldridge Company[1]

The DCS was introduced in 1975. Both Honeywell and Japanese electrical engineering firm Yokogawa introduced their own independently produced DCSs at roughly the same time, with the TDC 2000 and CENTUM[2] systems, respectively. US-based Bristol also introduced their UCS 3000 universal controller in 1975. In 1978 Metso(known as Valmet in 1978) introduced their own DCS system called Damatic (latest generation named Metso DNA[3]). In 1980, Bailey (now part of ABB[4]) introduced the NETWORK 90 system. Also in 1980, Fischer & Porter Company (now also part of ABB[5]) introduced DCI-4000 (DCI stands for Distributed Control Instrumentation).

The DCS largely came about due to the increased availability of microcomputers and the proliferation of microprocessors in the world of process control. Computers had already been applied to process automation for some time in the form of both direct digital control (DDC) and set point control. In the early 1970s Taylor Instrument Company, (now part of ABB) developed the 1010 system, Foxboro the FOX1 system and Bailey Controls the 1055 systems. All of these were DDC applications implemented within minicomputers (DEC PDP-11, Varian Data Machines, MODCOMP etc.) and connected to proprietary Input/Output hardware. Sophisticated (for the time) continuous as well as batch control was implemented in this way. A more conservative approach was set point control, where process computers supervised clusters of analog process controllers. A CRT-based workstation provided visibility into the process using text and crude character graphics. Availability of a fully functional graphical user interface was a way away.

Central to the DCS model was the inclusion of control function blocks. Function blocks evolved from early, more primitive DDC concepts of "Table Driven" software. One of the first embodiments of object-oriented software, function blocks were self-contained "blocks" of code that emulated analog hardware control components and performed tasks that were essential to process control, such as execution of PID algorithms. Function blocks continue to endure as the predominant method of control for DCS suppliers, and are supported by key technologies such as **Foundation Fieldbus**[6] today.

Midac Systems, of Sydney, Australia, developed an objected-oriented distributed direct digital control system in 1982. The central system ran 11 microprocessors sharing tasks and common memory and connected to a serial communication network of distributed controllers each running two Z80s. The system was installed at the University of Melbourne.

Digital communication between distributed controllers, workstations and other computing elements (peer to peer access) was one of the primary advantages of the DCS. Attention was duly focused on the networks, which provided the all-important lines of communication that, for process applications, had to incorporate specific functions such as determinism and redundancy. As a result, many suppliers embraced the IEEE 802.4 networking standard. This decision set the stage for the wave of migrations necessary when information technology moved into process automation and IEEE 802.3 rather than IEEE 802.4 prevailed as the control LAN.

The Network Centric Era of the 1980s

The DCS brought distributed intelligence to the plant and established the presence of computers and microprocessors in process control, but it still did not provide the reach and openness necessary to unify plant resource requirements. In many cases, the DCS was merely a digital replacement of the same functionality provided by analog controllers and a panelboard display. This was embodied in The Purdue Reference Model (PRM) that was developed to define Manufacturing Operations Management relationships. PRM later formed the basis for ISA-95 standards activities today.

In the 1980s, users began to look at DCSs as more than just basic process control. A very early example of a Direct Digital Control DCS was completed by the Australian business Midac in 1981–82 using R-Tec Australian designed hardware. The system installed at the University of Melbourne used a serial communications network, connecting campus buildings back to a control room "front end". Each remote unit ran 2 Z80 microprocessors whilst the front end ran 11 in a Parallel Processing configuration with paged common memory to share tasks and could run up to 20,000 concurrent controls objects.

It was believed that if openness could be achieved and greater amounts of data could be shared throughout the enterprise that even greater things could be achieved. The first attempts to increase the openness of DCSs resulted in the adoption of the predominant operating system of the day: *UNIX*. UNIX and its companion networking technology TCP-IP were developed by the US Department of Defense for openness, which was precisely the issue the process industries were looking to resolve.

As a result suppliers also began to adopt Ethernet-based networks with their own proprietary protocol layers. The full TCP/IP standard was not implemented, but the use of Ethernet made it possible to implement the first instances of object management and global data access technology. The 1980s also witnessed the first PLCs integrated into the DCS infrastructure. Plant-wide historians also emerged to capitalize on the extended reach of automation systems. The first DCS supplier to adopt UNIX and Ethernet networking technologies was Foxboro, who introduced the I/A Series[7] system in 1987.

The application-centric era of the 1990s

The drive toward openness in the 1980s gained momentum through the 1990s with the increased adoption of commercial off-the-shelf (COTS) components and IT standards. Probably the biggest transition undertaken during this time was the move from the UNIX operating system to the Windows environment. While the realm of the real time operating system (RTOS) for control applications remains dominated by real time commercial variants of UNIX or proprietary operating systems, everything above real-time control has made the transition to Windows.

The introduction of Microsoft at the desktop and server layers resulted in the development of technologies such as OLE for process control (OPC), which is now a de facto industry connectivity standard. Internet technology also began to make its mark in automation and the DCS world, with most DCS HMI supporting Internet connectivity. The 1990s were also known for the "Fieldbus Wars", where rival organizations competed to define what would become the IEC fieldbus standard for digital communication with field instrumentation instead of 4–20 milliamp analog communications. The first fieldbus installations occurred in the 1990s. Towards the end of the decade, the technology began to develop significant momentum, with the market consolidated around Ethernet I/P, Foundation Fieldbus and Profibus PA for process automation applications. Some suppliers built new systems from the ground up to maximize functionality with fieldbus, such as Rockwell PlantPAX System, Honeywell with Experion & Plantscape SCADA systems, ABB with System 800xA,[8] Emerson Process Management[9] with the Emerson Process Management DeltaV control system, Siemens with the SPPA-T3000 [10] or Simatic PCS 7[11] and **Azbil Corporation**[12] with the Harmonas-DEO system. Fieldbus technics have been used to integrate machine, drives, quality and condition monitoring applications to one DCS with Metso DNA system.[3]

The impact of COTS, however, was most pronounced at the hardware layer. For years, the primary business of DCS suppliers had been the supply of large amounts of hardware, particularly I/O and controllers. The initial proliferation of DCSs required the installation of prodigious amounts of this hardware, most of it manufactured from the bottom up by DCS suppliers. Standard computer components from manufacturers such as Intel and Motorola, however, made it cost prohibitive for DCS suppliers to continue making their own components, workstations, and networking hardware.

As the suppliers made the transition to COTS components, they also discovered that the hardware market was shrinking fast. COTS not only resulted in lower manufacturing costs for the supplier, but also steadily decreasing prices for the end users, who were also becoming increasingly vocal over what they perceived to be unduly high hardware costs. Some suppliers that were previously stronger in the PLC business, such as Rockwell Automation and Siemens, were able to leverage their expertise in manufacturing control hardware to enter the DCS marketplace with cost effective offerings, while the stability/scalability/reliability and functionality of these emerging systems are still improving. The traditional DCS suppliers introduced new generation DCS System based on the latest Communication and IEC Standards, which resulting in a trend of combining the traditional concepts/functionalities for PLC and DCS into a one for all solution—named "Process Automation System". The gaps among the various systems remain at the areas such as: the database integrity, pre-engineering functionality, system maturity, communication transparency and reliability. While it is expected the cost ratio is relatively the same (the more powerful the systems are, the more expensive they will be), the reality of the automation business is often operating strategically case by case. The current next evolution step is called Collaborative Process Automation Systems.

To compound the issue, suppliers were also realizing that the hardware market was becoming saturated. The lifecycle of hardware components such as I/O and wiring is also typically in the range of 15 to over 20 years, making for a challenging replacement market. Many of the older systems that were installed in the 1970s and 1980s are still in use today, and there is a considerable installed base of systems in the market that are approaching the end of their useful life. Developed industrial economies in North America, Europe, and Japan already had many thousands of DCSs installed, and with few if any new plants being built, the market for new hardware was shifting rapidly to smaller, albeit faster growing regions such as China, Latin America, and Eastern Europe.

Because of the shrinking hardware business, suppliers began to make the challenging transition from a hardware-based business model to one based on software and value-added services. It is a transition that is still being made today. The applications portfolio offered by suppliers expanded considerably in the '90s to include areas such as production management, model-based control, real-time optimization, plant asset management (PAM), Real-time performance management (RPM) tools, alarm management, and many others. To obtain the true value from these applications, however, often requires a considerable service content, which the suppliers also provide.

See also

- OSIsoft
- Building Automation
- Direct Digital Control
- SCADA
- PLC
- Fieldbus
- First-out alarm
- Midac
- Safety instrumented system, (SIS)
- Industrial control systems
- Industrial safety systems
- Annunciator panel
- EPICS

References

[1] Stout, T. M. and Williams, T. J. (1995). "Pioneering Work in the Field of Computer Process Control". *IEEE Annals of the History of Computing* **17** (1).
[2] (http://www.yokogawa.com/dcs/dcs-index-en.htm) CENTUM
[3] (http://www.metso.com/metsodna) Metso DNA
[4] (http://www.abb.com/controlsystems) INFI 90
[5] (http://www.abb.com/product/us/9AAC115762.aspx) DCI-4000
[6] (http://www.fieldbus.org) Foundation Fieldbus
[7] (http://iom.invensys.com/UK/Pages/Foxboro_DCSIASeries.aspx) Foxboro I/A Series Distributed Control System
[8] ABB System 800xA (http://www.abb.com/product/us/9AAC115756.aspx)
[9] (http://easydeltav.com) Emerson Process Management
[10] (http://www.energy.siemens.com/hq/en/automation/power-generation/sppa-t3000.htm) SPPA-T3000
[11] (http://pcs.khe.siemens.com/index.aspx?nr=1075) Simatic PCS 7
[12] (http://www.azbil.com) Azbil Corporation

External links

- DCS Selection MBA research program with many Links (http://www.DCSSELECT.EU)
- Example of DCS system: Mark VIe by General Electric (http://www.gepower.com/prod_serv/products/oc/en/control_solution/ppc_markviedcs_cs.htm)
- An even better example of DCS system: Control Design Platform by ICD (http://www.icd.no/)
- Proview is probably the first Open Source system for process control and automation in the world. (http://www.proview.se/v2/)
- Open Source DCS and automation software (http://www.openapc.com/)
- FreeDCS is another Open Source Distributed Control System (http://freedcs.sourceforge.net/index.html/)

SCADA

SCADA (**supervisory control and data acquisition**) is a type of industrial control system (ICS). Industrial control systems are computer controlled systems that monitor and control industrial processes that exist in the physical world. SCADA systems historically distinguish themselves from other ICS systems by being large scale processes that can include multiple sites, and large distances.[1] These processes include industrial, infrastructure, and facility-based processes, as described below:

- Industrial processes include those of manufacturing, production, power generation, fabrication, and refining, and may run in continuous, batch, repetitive, or discrete modes.
- Infrastructure processes may be public or private, and include water treatment and distribution, wastewater collection and treatment, oil and gas pipelines, electrical power transmission and distribution, wind farms, civil defense siren systems, and large communication systems.
- Facility processes occur both in public facilities and private ones, including buildings, airports, ships, and space stations. They monitor and control heating, ventilation, and air conditioning systems (HVAC), access, and energy consumption.

Common system components

A SCADA system usually consists of the following subsystems:

- A human–machine interface or HMI is the apparatus or device which presents process data to a human operator, and through this, the human operator monitors and controls the process.
- SCADA is used as a safety tool as in lock-out tag-out
- A supervisory (computer) system, gathering (acquiring) data on the process and sending commands (control) to the process.
- Remote terminal units (RTUs) connecting to sensors in the process, converting sensor signals to digital data and sending digital data to the supervisory system.
- Programmable logic controller (PLCs) used as field devices because they are more economical, versatile, flexible, and configurable than special-purpose RTUs.
- Communication infrastructure connecting the supervisory system to the remote terminal units.
- Various process and analytical instrumentation

Systems concepts

The term SCADA usually refers to centralized systems which monitor and control entire sites, or complexes of systems spread out over large areas (anything from an industrial plant to a nation). Most control actions are performed automatically by RTUs or by PLCs. Host control functions are usually restricted to basic overriding or *supervisory* level intervention. For example, a PLC may control the flow of cooling water through part of an industrial process, but the SCADA system may allow operators to change the set points for the flow, and enable alarm conditions, such as loss of flow and high temperature, to be displayed and recorded. The feedback control loop passes through the RTU or PLC, while the SCADA system monitors the overall performance of the loop.

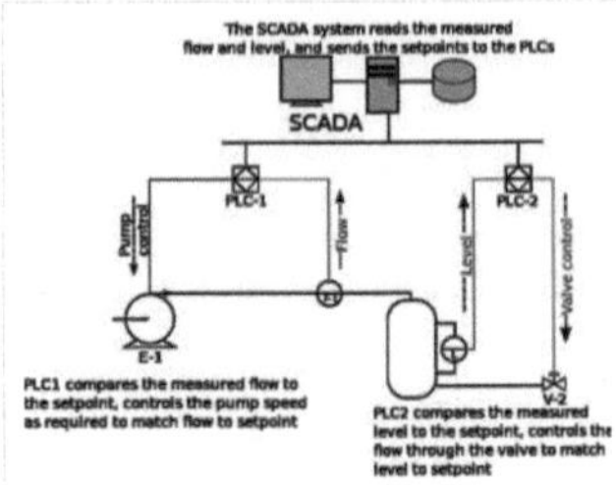

SCADA schematic overview

Data acquisition begins at the RTU or PLC level and includes meter readings and equipment status reports that are communicated to SCADA as required. Data is then compiled and formatted in such a way that a control room operator using the HMI can make supervisory decisions to adjust or override normal RTU (PLC) controls. Data may also be fed to an Historian, often built on a commodity Database Management System, to allow trending and other analytical auditing.

SCADA systems typically implement a distributed database, commonly referred to as a *tag database*, which contains data elements called *tags* or *points*. A point represents a single input or output value monitored or controlled by the system. Points can be either "hard" or "soft". A hard point represents an actual input or output within the system, while a soft point results from logic and math operations applied to other points. (Most implementations conceptually remove the distinction by making every property a "soft" point expression, which may, in the simplest case, equal a single hard point.) Points are normally stored as value-timestamp pairs: a value, and the timestamp when it was recorded or calculated. A series of value-timestamp pairs gives the history of that point. It is also common to store additional metadata with tags, such as the path to a field device or PLC register, design time comments, and alarm information.

SCADA systems are significantly important systems used in national infrastructures such as electric grids, water supplies and pipelines. However, SCADA systems have many security vulnerabilities.

Human–machine interface

A human–machine interface or HMI is the apparatus which presents process data to a human operator, and through which the human operator controls the process.

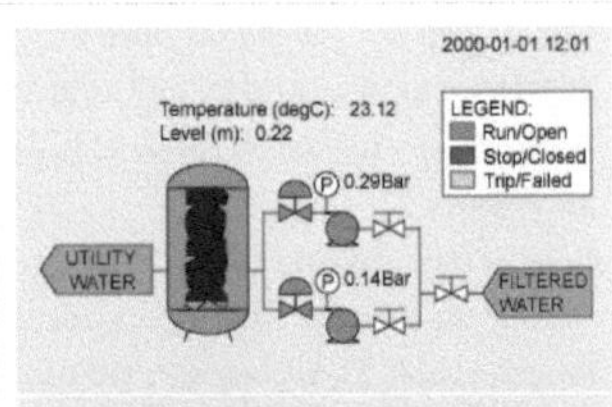

Typical basic SCADA animations [2]

HMI is usually linked to the SCADA system's databases and software programs, to provide trending, diagnostic data, and management information such as scheduled maintenance procedures, logistic information, detailed schematics for a particular sensor or machine, and expert-system troubleshooting guides.

The HMI system usually presents the information to the operating personnel graphically, in the form of a mimic diagram. This means that the operator can see a schematic representation of the plant being controlled. For example, a picture of a pump connected to a pipe can show the operator that the pump is running and how much fluid it is pumping through the pipe at the moment. The operator can then switch the pump off. The HMI software will show the flow rate of the fluid in the pipe decrease in real time. Mimic diagrams may consist of line graphics and schematic symbols to represent process elements, or may consist of digital photographs of the process equipment overlain with animated symbols.

The HMI package for the SCADA system typically includes a drawing program that the operators or system maintenance personnel use to change the way these points are represented in the interface. These representations can be as simple as an on-screen traffic light, which represents the state of an actual traffic light in the field, or as complex as a multi-projector display representing the position of all of the elevators in a skyscraper or all of the trains on a railway.

An important part of most SCADA implementations is alarm handling. The system monitors whether certain alarm conditions are satisfied, to determine when an alarm event has occurred. Once an alarm event has been detected, one or more actions are taken (such as the activation of one or more alarm indicators, and perhaps the generation of email or text messages so that management or remote SCADA operators are informed). In many cases, a SCADA operator may have to acknowledge the alarm event; this may deactivate some alarm indicators, whereas other indicators remain active until the alarm conditions are cleared. Alarm conditions can be explicit—for example, an alarm point is a digital status point that has either the value NORMAL or ALARM that is calculated by a formula based on the values in other analogue and digital points—or implicit: the SCADA system might automatically monitor whether the value in an analogue point lies outside high and low limit values associated with that point. Examples of alarm indicators include a siren, a pop-up box on a screen, or a coloured or flashing area on a screen (that might act in a similar way to the "fuel tank empty" light in a car); in each case, the role of the alarm indicator is to draw the operator's attention to the part of the system 'in alarm' so that appropriate action can be taken. In designing SCADA systems, care must be taken when a cascade of alarm events occurs in a short time, otherwise the underlying cause (which might not be the earliest event detected) may get lost in the noise. Unfortunately, when used as a noun, the word 'alarm' is used rather loosely in the industry; thus, depending on context it might mean an alarm point, an alarm indicator, or an alarm event.

Hardware solutions

SCADA solutions often have distributed control system (DCS) components. Use of "smart" RTUs or PLCs, which are capable of autonomously executing simple logic processes without involving the master computer, is increasing. A standardized control programming language, IEC 61131-3 (a suite of 5 programming languages including Function Block, Ladder, Structured Text, Sequence Function Charts and Instruction List), is frequently used to create programs which run on these RTUs and PLCs. Unlike a procedural language such as the C programming language or FORTRAN, IEC 61131-3 has minimal training requirements by virtue of resembling historic physical control arrays. This allows SCADA system engineers to perform both the design and implementation of a program to be executed on an RTU or PLC. A programmable automation controller (PAC) is a compact controller that combines the features and capabilities of a PC-based control system with that of a typical PLC. PACs are deployed in SCADA systems to provide RTU and PLC functions. In many electrical substation SCADA applications, "distributed RTUs" use information processors or station computers to communicate with digital protective relays, PACs, and other devices for I/O, and communicate with the SCADA master in lieu of a traditional RTU.

Since about 1998, virtually all major PLC manufacturers have offered integrated HMI/SCADA systems, many of them using open and non-proprietary communications protocols. Numerous specialized third-party HMI/SCADA packages, offering built-in compatibility with most major PLCs, have also entered the market, allowing mechanical engineers, electrical engineers and technicians to configure HMIs themselves, without the need for a custom-made program written by a software developer.

Remote terminal unit

The remote terminal unit(RTU) connects to physical equipment. Typically, an RTU converts the electrical signals from the equipment to digital values such as the open/closed status from a switch or a valve, or measurements such as pressure, flow, voltage or current. By converting and sending these electrical signals out to equipment the RTU can control equipment, such as opening or closing a switch or a valve, or setting the speed of a pump.

Supervisory station

The term *supervisory station* refers to the servers and software responsible for communicating with the field equipment (RTUs, PLCs, etc.), and then to the HMI software running on workstations in the control room, or elsewhere. In smaller SCADA systems, the master station may be composed of a single PC. In larger SCADA systems, the master station may include multiple servers, distributed software applications, and disaster recovery sites. To increase the integrity of the system the multiple servers will often be configured in a dual-redundant or hot-standby formation providing continuous control and monitoring in the event of a server failure.

Operational philosophy

For some installations, the costs that would result from the control system failing are extremely high. Possibly even lives could be lost. Hardware for some SCADA systems is ruggedized to withstand temperature, vibration, and voltage extremes. In the most critical installations, reliability is enhanced by having redundant hardware and communications channels, up to the point of having multiple fully equipped control centres. A failing part can be quickly identified and its functionality automatically taken over by backup hardware. A failed part can often be replaced without interrupting the process. The reliability of such systems can be calculated statistically and is stated as the mean time to failure, which is a variant of mean time between failures. The calculated mean time to failure of such high reliability systems can be on the order of centuries.

Communication infrastructure and methods

SCADA systems have traditionally used combinations of radio and direct wired connections, although SONET / SDH is also frequently used for large systems such as railways and power stations. The remote management or monitoring function of a SCADA system is often referred to as telemetry. Some users want SCADA data to travel over their pre-established corporate networks or to share the network with other applications. The legacy of the early low-bandwidth protocols remains, though.

SCADA protocols are designed to be very compact. Many are designed to send information only when the master station polls the RTU. Typical legacy SCADA protocols include Modbus RTU, RP-570, Profibus and Conitel. These communication protocols are all SCADA-vendor specific but are widely adopted and used. Standard protocols are IEC 60870-5-101 or 104, IEC 61850 and DNP3. These communication protocols are standardized and recognized by all major SCADA vendors. Many of these protocols now contain extensions to operate over TCP/IP.

With increasing security demands (such as North American Electric Reliability Corporation (NERC) and critical infrastructure protection (CIP) in the US), there is increasing use of satellite-based communication. This has the key advantages that the infrastructure can be self contained (not using circuits from the public telephone system), can have built-in encryption, and can be engineered to the availability and reliability required by the SCADA system operator. Earlier experiences using consumer-grade VSAT were poor. Modern carrier-class systems provide the quality of service required for SCADA.[3]

RTUs and other automatic controller devices were developed before the advent of industry wide standards for interoperability. The result is that developers and their management created a multitude of control protocols. Among the larger vendors, there was also the incentive to create their own protocol to "lock in" their customer base. A list of automation protocols is compiled here.

Recently, OLE for process control (OPC) has become a widely accepted solution for intercommunicating different hardware and software, allowing communication even between devices originally not intended to be part of an industrial network.

SCADA architectures

SCADA systems have evolved through 3 generations as follows:[4]

First generation: "Monolithic"

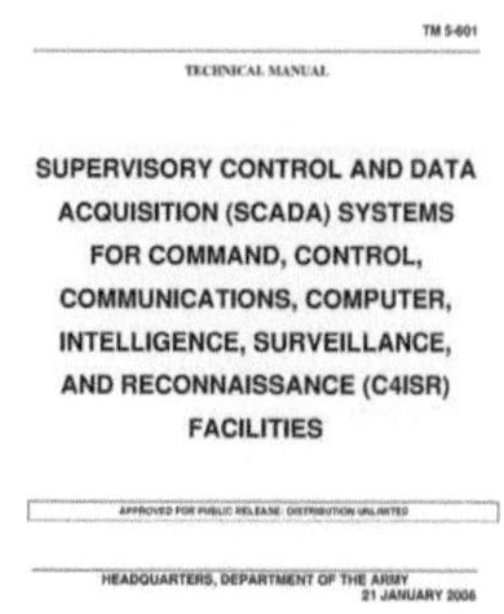

TM 5-601

TECHNICAL MANUAL

SUPERVISORY CONTROL AND DATA ACQUISITION (SCADA) SYSTEMS FOR COMMAND, CONTROL, COMMUNICATIONS, COMPUTER, INTELLIGENCE, SURVEILLANCE, AND RECONNAISSANCE (C4ISR) FACILITIES

APPROVED FOR PUBLIC RELEASE: DISTRIBUTION UNLIMITED

HEADQUARTERS, DEPARTMENT OF THE ARMY
21 JANUARY 2006

The United States Army's Training Manual 5-601 covers "SCADA Systems for C4ISR Facilities".

In the first generation, computing was done by mainframe computers. Networks did not exist at the time SCADA was developed. Thus SCADA systems were independent systems with no connectivity to other systems. Wide Area Networks were later designed by RTU vendors to communicate with the RTU. The communication protocols used were often proprietary at that time. The first-generation SCADA system was redundant since a back-up mainframe system was connected at the bus level and was used in the event of failure of the primary mainframe system. Some first generation SCADA systems were developed as "turn key" operations that ran on minicomputers like the PDP-11 series made by the Digital Equiptment Corporation (DEC). These systems were read only in the sense that they could display information from the existing analog based control systems to individual operator workstations but they usually didn't attempt to send control signals to remote stations due to anaolog based telemetry issues and control center management concerns with allowing direct control from computer workstations. They would also perform alarming and logging functions and calculate hourly and daily system commodity accounting functions.

Second generation: "Distributed"

The processing was distributed across multiple stations which were connected through a LAN and they shared information in real time. Each station was responsible for a particular task thus making the size and cost of each station less than the one used in First Generation. The network protocols used were still mostly proprietary, which led to significant security problems for any SCADA system that received attention from a hacker. Since the protocols were proprietary, very few people beyond the developers and hackers knew enough to determine how secure a SCADA installation was. Since both parties had vested interests in keeping security issues quiet, the security of a SCADA installation was often badly overestimated, if it was considered at all.

Third generation: "Networked"

Due to the usage of standard protocols and the fact that many networked SCADA systems are accessible from the Internet, the systems are potentially vulnerable to remote attack. On the other hand, the usage of standard protocols and security techniques means that standard security improvements are applicable to the SCADA systems, assuming they receive timely maintenance and updates

Security issues

The move from proprietary technologies to more standardized and open solutions together with the increased number of connections between SCADA systems and office networks and the Internet has made them more vulnerable to attacks—see references. Consequently, the security of some SCADA-based systems has come into question as they are seen as potentially vulnerable to cyber attacks.[5] [6] [7]

In particular, security researchers are concerned about:

- the lack of concern about security and authentication in the design, deployment and operation of some existing SCADA networks
- the belief that SCADA systems have the benefit of security through obscurity through the use of specialized protocols and proprietary interfaces
- the belief that SCADA networks are secure because they are physically secured
- the belief that SCADA networks are secure because they are disconnected from the Internet.

SCADA systems are used to control and monitor physical processes, examples of which are transmission of electricity, transportation of gas and oil in pipelines, water distribution, traffic lights, and other systems used as the basis of modern society. The security of these SCADA systems is important because compromise or destruction of these systems would impact multiple areas of society far removed from the original compromise. For example, a blackout caused by a compromised electrical SCADA system would cause financial losses to all the customers that received electricity from that source. How security will affect legacy SCADA and new deployments remains to be seen.

There are two distinct threats to a modern SCADA system. First is the threat of unauthorized access to the control software, whether it be human access or changes induced intentionally or accidentally by virus infections and other software threats residing on the control host machine. Second is the threat of packet access to the network segments hosting SCADA devices. In many cases, there is rudimentary or no security on the actual packet control protocol, so anyone who can send packets to the SCADA device can control it. In many cases SCADA users assume that a VPN is sufficient protection and are unaware that physical access to SCADA-related network jacks and switches provides the ability to totally bypass all security on the control software and fully control those SCADA networks. These kinds of physical access attacks bypass firewall and VPN security and are best addressed by endpoint-to-endpoint authentication and authorization such as are commonly provided in the non-SCADA world by in-device SSL or other cryptographic techniques.

The reliable function of SCADA systems in our modern infrastructure may be crucial to public health and safety. As such, attacks on these systems may directly or indirectly threaten public health and safety. Such an attack has already occurred, carried out on Maroochy Shire Council's sewage control system in Queensland, Australia.[8] Shortly after a contractor installed a SCADA system in January 2000, system components began to function erratically. Pumps did not run when needed and alarms were not reported. More critically, sewage flooded a nearby park and contaminated an open surface-water drainage ditch and flowed 500 meters to a tidal canal. The SCADA system was directing sewage valves to open when the design protocol should have kept them closed. Initially this was believed to be a system bug. Monitoring of the system logs revealed the malfunctions were the result of cyber attacks. Investigators reported 46 separate instances of malicious outside interference before the culprit was identified. The attacks were made by a disgruntled employee of the company that had installed the SCADA system. The employee was hoping to be hired full time to help solve the problem.

Many vendors of SCADA and control products have begun to address the risks posed by unauthorized access by developing lines of specialized industrial firewall and VPN solutions for TCP/IP-based SCADA networks as well as external SCADA monitoring and recording equipment.[9] The International Society of Automation (ISA) started formalizing SCADA security requirements in 2007 with a working group, WG4. WG4 "deals specifically with unique technical requirements, measurements, and other features required to evaluate and assure security resilience and performance of industrial automation and control systems devices".[10]

The increased interest in SCADA vulnerabilities has resulted in vulnerability researchers discovering vulnerabilities in commercial SCADA software and more general offensive SCADA techniques presented to the general security community.[11] [12] In electric and gas utility SCADA systems, the vulnerability of the large installed base of wired and wireless serial communications links is addressed in some cases by applying bump-in-the-wire devices that employ authentication and Advanced Encryption Standard encryption rather than replacing all existing nodes.[13]

In June 2010, anti-virus security company VirusBlokAda reported the first detection of malware that attacks SCADA systems (Siemens' WinCC/PCS 7 systems) running on Windows operating systems. The malware is called Stuxnet and uses four zero-day attacks to install a rootkit which in turn logs in to the SCADA's database and steals design and control files.[14] [15] The malware is also capable of changing the control system and hiding those changes. The malware was found on 14 systems, the majority of which were located in Iran.[16]

See also

- BACnet
- LonWorks
- Modbus
- Telemetry
- EPICS

References

[1] Boyes, Walt (18 August 2009). "Back to Basics: SCADA" (http://www.youtube.com/watch?NR=1&v=bfxr5DikdP0). Automation TV: Control Global - Control Design. .

[2] Basic SCADA Animations (http://www.integraxor.com/en/index.htm?sl=wk&menu=intro)

[3] Bergan, Christian (August 2011). "Demystifying Satellite for the Smart Grid: Four Common Misconceptions" (http://www.elp.com/index/display/article-display/5666163079/articles/utility-automation-engineering-td/volume-16/issue-8/features/demystifying-satellite-for-the-smart-grid-four-common-misconceptions.html). *Electric Light & Power*. Utility Automation & Engineering T&D (Tulsa, OK: PennWell) **16** (8): Four. . Retrieved 2 May 2012. "satellite is a cost-effective and secure solution that can provide backup communications and easily support core smart grid applications like SCADA, telemetry, AMI backhaul and distribution automation"

[4] OFFICE OF THE MANAGER NATIONAL COMMUNICATIONS SYSTEM (October 2004). "Supervisory Control and Data Acquisition (SCADA) Systems" (http://www.ncs.gov/library/tech_bulletins/2004/tib_04-1.pdf). NATIONAL COMMUNICATIONS SYSTEM. .

[5] D. Maynor and R. Graham (2006). "SCADA Security and Terrorism: We're Not Crying Wolf" (http://www.blackhat.com/presentations/bh-federal-06/BH-Fed-06-Maynor-Graham-up.pdf). .

[6] Robert Lemos (26 July 2006). "SCADA system makers pushed toward security" (http://www.securityfocus.com/news/11402). SecurityFocus. . Retrieved 9 May 2007.

[7] "Cyberthreats, Vulnerabilities and Attacks on SCADA Networks" (http://gspp.berkeley.edu/iths/Tsang_SCADA Attacks.pdf). *Rosa Tang, berkeley.edu*. . Retrieved 1 August 2012.

[8] Slay, J.; M. Miller (November 2007). "Chpt 6: Lessons Learned from the Maroochy Water Breach" (http://www.springerlink.com/content/m1100j53q348x053/). *Critical infrastructure protection* (Online-Ausg. ed.). Springer Boston. pp. 73–82. ISBN 978-0-387-75461-1. . Retrieved 2 May 2012.

[9] "External SCADA Monitoring" (http://www.epiphan.com/solutions_new/?arid=84). *Epiphan Case Studies*. Epiphan Systems Inc.. . Retrieved 2 May 2012.

[10] "Security for all" (http://www.isa.org). *InTech*. June 2008. . Retrieved 2 May 2012.

[11] "S4 2008 Agenda" (http://www.digitalbond.com/wp-content/uploads/2007/10/S4_2008_Agenda.pdf). .

[12] "SCADA Security – Generic Electric Grid Malware Design" (http://www.c4-security.com/SCADA Security - Generic Electric Grid Malware Design - SyScan08.pps). .

[13] KEMA, Inc. (November 2006). *Substation Communications: Enabler of Automation / An Assessment of Communications Technologies*. UTC – United Telecom Council. pp. 3–21.

[14] Mills, Elinor (21 July 2010). "Details of the first-ever control system malware (FAQ)" (http://news.cnet.com/8301-27080_3-20011159-245.html). CNET. . Retrieved 21 July 2010.

[15] "SIMATIC WinCC / SIMATIC PCS 7: Information concerning Malware / Virus / Trojan" (http://support.automation.siemens.com/WW/llisapi.dll?func=cslib.csinfo&lang=en&objid=43876783&caller=view). Siemens. 21 July 2010. . Retrieved 22 July 2010. "malware (trojan) which affects the visualization system WinCC SCADA."

[16] "Siemens: Stuxnet worm hit industrial systems" (http://www.computerworld.com/s/article/print/9185419/Siemens_Stuxnet_worm_hit_industrial_systems?taxonomyName=Network+Security&taxonomyId=142). . Retrieved 16 September 2010.

External links

- UK SCADA security guidelines (http://www.cpni.gov.uk/advice/infosec/business-systems/scada/)
- BBC NEWS | Technology | Spies 'infiltrate US power grid' (http://news.bbc.co.uk/1/hi/technology/7990997.stm)
- Back to Basics: SCADA (YouTube video sponsored by Schneider Electric) (http://www.youtube.com/watch?NR=1&v=bfxr5DikdP0)

Programmable_logic_controller

A **Programmable Logic Controller**, PLC or **Programmable Controller** is a digital computer used for automation of electromechanical processes, such as control of machinery on factory assembly lines, amusement rides, or light fixtures. PLCs are used in many industries and machines. Unlike general-purpose computers, the PLC is designed for multiple inputs and output arrangements, extended temperature ranges, immunity to electrical noise, and resistance to vibration and impact. Programs to control machine operation are typically stored in battery-backed-up or non-volatile memory. A PLC is an example of a *hard* real time system since output results must be produced in response to input conditions within a limited time, otherwise unintended operation will result.

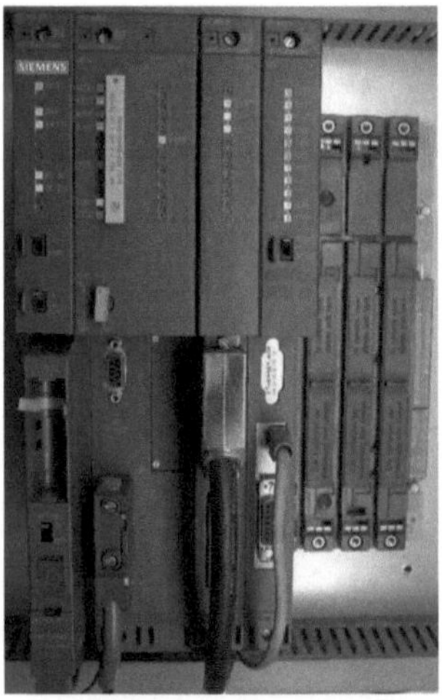

Siemens Simatic S7-400 system at rack, left-to-right: power supply unit PS407 4A,CPU 416-3, interface module IM 460-0 and communication processor CP 443-1.

History

Before the PLC, control, sequencing, and safety interlock logic for manufacturing automobiles was mainly composed of relays, cam timers, drum sequencers, and dedicated closed-loop controllers. Since these could number in the hundreds or even thousands, the process for updating such facilities for the yearly model change-over was very time consuming and expensive, as electricians needed to individually rewire relays to change the logic.

Digital computers, being general-purpose programmable devices, were soon applied to control of industrial processes. Early computers required specialist programmers, and stringent operating environmental control for temperature, cleanliness, and power quality. Using a general-purpose computer for process control required protecting the computer from the plant floor conditions. An industrial control computer would have several attributes: it would tolerate the shop-floor environment, it would support discrete (bit-form) input and output in an easily extensible manner, it would not require years of training to use, and it would permit its operation to be monitored. The response time of any computer system must be fast enough to be useful for control; the required speed varying according to the nature of the process.[1]

In 1968 GM Hydramatic (the automatic transmission division of General Motors) issued a request for proposal for an electronic replacement for hard-wired relay systems. The winning proposal came from Bedford Associates of Bedford, Massachusetts. The first PLC, designated the 084 because it was Bedford Associates' eighty-fourth project, was the result.[2] Bedford Associates started a new company dedicated to developing, manufacturing, selling, and servicing this new product: Modicon, which stood for **MO**dular **DI**gital **CON**troller. One of the people who worked on that project was Dick Morley, who is considered to be the "father" of the PLC.[3] The Modicon brand was sold in 1977 to Gould Electronics, and later acquired by German Company AEG and then by French Schneider Electric, the current owner.

One of the very first 084 models built is now on display at Modicon's headquarters in North Andover, Massachusetts. It was presented to Modicon by GM, when the unit was retired after nearly twenty years of uninterrupted service. Modicon used the 84 moniker at the end of its product range until the 984 made its appearance.

The automotive industry is still one of the largest users of PLCs.

Development

Early PLCs were designed to replace relay logic systems. These PLCs were programmed in "ladder logic", which strongly resembles a schematic diagram of relay logic. This program notation was chosen to reduce training demands for the existing technicians. Other early PLCs used a form of instruction list programming, based on a stack-based logic solver.

Modern PLCs can be programmed in a variety of ways, from the relay-derived ladder logic to programming languages such as specially adapted dialects of BASIC and C. Another method is State Logic, a very high-level programming language designed to program PLCs based on state transition diagrams.

Many early PLCs did not have accompanying programming terminals that were capable of graphical representation of the logic, and so the logic was instead represented as a series of logic expressions in some version of Boolean format, similar to Boolean algebra. As programming terminals evolved, it became more common for ladder logic to be used, for the aforementioned reasons and because it was a familiar format used for electromechanical control panels. Newer formats such as State Logic and Function Block (which is similar to the way logic is depicted when using digital integrated logic circuits) exist, but they are still not as popular as ladder logic. A primary reason for this is that PLCs solve the logic in a predictable and repeating sequence, and ladder logic allows the programmer (the person writing the logic) to see any issues with the timing of the logic sequence more easily than would be possible in other formats.

Programming

Early PLCs, up to the mid-1980s, were programmed using proprietary programming panels or special-purpose programming terminals, which often had dedicated function keys representing the various logical elements of PLC programs.[2] Programs were stored on cassette tape cartridges. Facilities for printing and documentation were minimal due to lack of memory capacity. The very oldest PLCs used non-volatile magnetic core memory.

More recently, PLCs are programmed using application software on personal computers. The computer is connected to the PLC through Ethernet, RS-232, RS-485 or RS-422 cabling. The programming software allows entry and editing of the ladder-style logic. Generally the software provides functions for debugging and troubleshooting the PLC software, for example, by highlighting portions of the logic to show current status during operation or via simulation. The software will upload and download the PLC program, for backup and restoration purposes. In some models of programmable controller, the program is transferred from a personal computer to the PLC through a programming board which writes the program into a removable chip such as an EEPROM or EPROM.

Functionality

The functionality of the PLC has evolved over the years to include sequential relay control, motion control, process control, distributed control systems and networking. The data handling, storage, processing power and communication capabilities of some modern PLCs are approximately equivalent to desktop computers. PLC-like programming combined with remote I/O hardware, allow a general-purpose desktop computer to overlap some PLCs in certain applications. Regarding the practicality of these desktop computer based logic controllers, it is important to note that they have not been generally accepted in heavy industry because the desktop computers run on less stable operating systems than do PLCs, and because the desktop computer hardware is typically not designed to the same levels of tolerance to temperature, humidity, vibration, and longevity as the processors used in PLCs. In addition to the hardware limitations of desktop based logic, operating systems such as Windows do not lend themselves to deterministic logic execution, with the result that the logic may not always respond to changes in logic state or input status with the extreme consistency in timing as is expected from PLCs. Still, such desktop logic applications find use in less critical situations, such as laboratory automation and use in small facilities where the application is less demanding and critical, because they are generally much less expensive than PLCs.

Programmable Logic Relay (PLR)

In more recent years, small products called PLRs (programmable logic relays), and also by similar names, have become more common and accepted. These are very much like PLCs, and are used in light industry where only a few points of I/O (i.e. a few signals coming in from the real world and a few going out) are involved, and low cost is desired. These small devices are typically made in a common physical size and shape by several manufacturers, and branded by the makers of larger PLCs to fill out their low end product range. Popular names include PICO Controller, NANO PLC, and other names implying very small controllers. Most of these have between 8 and 12 discrete inputs, 4 and 8 discrete outputs, and up to 2 analog inputs. Size is usually about 4" wide, 3" high, and 3" deep. Most such devices include a tiny postage stamp sized LCD screen for viewing simplified ladder logic (only a very small portion of the program being visible at a given time) and status of I/O points, and typically these screens are accompanied by a 4-way rocker push-button plus four more separate push-buttons, similar to the key buttons on a VCR remote control, and used to navigate and edit the logic. Most have a small plug for connecting via RS-232 or RS-485 to a personal computer so that programmers can use simple Windows applications for programming instead of being forced to use the tiny LCD and push-button set for this purpose. Unlike regular PLCs that are usually modular and greatly expandable, the PLRs are usually not modular or expandable, but their price can be two orders of magnitude less than a PLC and they still offer robust design and deterministic execution of the logic.

PLC topics

Features

The main difference from other computers is that PLCs are armored for severe conditions (such as dust, moisture, heat, cold) and have the facility for extensive input/output (I/O) arrangements. These connect the PLC to sensors and actuators. PLCs read limit switches, analog process variables (such as temperature and pressure), and the positions of complex positioning systems. Some use machine vision.[4] On the actuator side, PLCs operate electric motors, pneumatic or hydraulic cylinders, magnetic relays, solenoids, or analog outputs. The input/output arrangements may be built into a simple PLC, or the PLC may have external I/O modules attached to a computer network that plugs into the PLC.

Control panel with PLC (grey elements in the center). The unit consists of separate elements, from left to right; power supply, controller, relay units for in- and output

Scan time

A PLC program is generally executed repeatedly as long as the controlled system is running. The status of physical input points is copied to an area of memory accessible to the processor, sometimes called the "I/O Image Table". The program is then run from its first instruction rung down to the last rung. It takes some time for the processor of the PLC to evaluate all the rungs and update the I/O image table with the status of outputs.[5] This scan time may be a few milliseconds for a small program or on a fast processor, but older PLCs running very large programs could take much longer (say, up to 100 ms) to execute the program. If the scan time was too long, the response of the PLC to process conditions would be too slow to be useful.

As PLCs became more advanced, methods were developed to change the sequence of ladder execution, and subroutines were implemented.[6] This simplified programming and could also be used to save scan time for high-speed processes; for example, parts of the program used only for setting up the machine could be segregated from those parts required to operate at higher speed.

Special-purpose I/O modules, such as timer modules or counter modules, can be used where the scan time of the processor is too long to reliably pick up, for example, counting pulses and interpreting quadrature from a shaft encoder. The relatively slow PLC can still interpret the counted values to control a machine, but the accumulation of pulses is done by a dedicated module that is unaffected by the speed of the program execution.

System scale

A small PLC will have a fixed number of connections built in for inputs and outputs. Typically, expansions are available if the base model has insufficient I/O.

Modular PLCs have a chassis (also called a rack) into which are placed modules with different functions. The processor and selection of I/O modules are customized for the particular application. Several racks can be administered by a single processor, and may have thousands of inputs and outputs. A special high speed serial I/O link is used so that racks can be distributed away from the processor, reducing the wiring costs for large plants.

User interface

PLCs may need to interact with people for the purpose of configuration, alarm reporting or everyday control. A human-machine interface (HMI) is employed for this purpose. HMIs are also referred to as man-machine interfaces (MMIs) and graphical user interface (GUIs). A simple system may use buttons and lights to interact with the user. Text displays are available as well as graphical touch screens. More complex systems use programming and monitoring software installed on a computer, with the PLC connected via a communication interface.

Communications

PLCs have built in communications ports, usually 9-pin RS-232, but optionally EIA-485 or Ethernet. Modbus, BACnet or DF1 is usually included as one of the communications protocols. Other options include various fieldbuses such as DeviceNet or Profibus. Other communications protocols that may be used are listed in the List of automation protocols.

Most modern PLCs can communicate over a network to some other system, such as a computer running a SCADA (Supervisory Control And Data Acquisition) system or web browser.

PLCs used in larger I/O systems may have peer-to-peer (P2P) communication between processors. This allows separate parts of a complex process to have individual control while allowing the subsystems to co-ordinate over the communication link. These communication links are also often used for HMI devices such as keypads or PC-type workstations.

Programming

PLC programs are typically written in a special application on a personal computer, then downloaded by a direct-connection cable or over a network to the PLC. The program is stored in the PLC either in battery-backed-up RAM or some other non-volatile flash memory. Often, a single PLC can be programmed to replace thousands of relays.[7]

Under the IEC 61131-3 standard, PLCs can be programmed using standards-based programming languages. A graphical programming notation called Sequential Function Charts is available on certain programmable controllers. Initially most PLCs utilized Ladder Logic Diagram Programming, a model which emulated electromechanical control panel devices (such as the contact and coils of relays) which PLCs replaced. This model remains common today.

IEC 61131-3 currently defines five programming languages for programmable control systems: function block diagram (FBD), ladder diagram (LD), structured text (ST; similar to the Pascal programming language), instruction list (IL; similar to assembly language) and sequential function chart (SFC).[8] These techniques emphasize logical organization of operations.[7]

While the fundamental concepts of PLC programming are common to all manufacturers, differences in I/O addressing, memory organization and instruction sets mean that PLC programs are never perfectly interchangeable between different makers. Even within the same product line of a single manufacturer, different models may not be directly compatible.

PLC compared with other control systems

Allen-Bradley PLC installed in a control panel

PLCs are well adapted to a range of automation tasks. These are typically industrial processes in manufacturing where the cost of developing and maintaining the automation system is high relative to the total cost of the automation, and where changes to the system would be expected during its operational life. PLCs contain input and output devices compatible with industrial pilot devices and controls; little electrical design is required, and the design problem centers on expressing the desired sequence of operations. PLC applications are typically highly customized systems, so the cost of a packaged PLC is low compared to the cost of a specific custom-built controller design. On the other hand, in the case of mass-produced goods, customized control systems are economical. This is due to the lower cost of the components, which can be optimally chosen instead of a "generic" solution, and where the non-recurring engineering charges are spread over thousands or millions of units.

For high volume or very simple fixed automation tasks, different techniques are used. For example, a consumer dishwasher would be controlled by an electromechanical cam timer costing only a few dollars in production quantities.

A microcontroller-based design would be appropriate where hundreds or thousands of units will be produced and so the development cost (design of power supplies, input/output hardware and necessary testing and certification) can be spread over many sales, and where the end-user would not need to alter the control. Automotive applications are an example; millions of units are built each year, and very few end-users alter the programming of these controllers. However, some specialty vehicles such as transit buses economically use PLCs instead of custom-designed controls, because the volumes are low and the development cost would be uneconomical.[9]

Very complex process control, such as used in the chemical industry, may require algorithms and performance beyond the capability of even high-performance PLCs. Very high-speed or precision controls may also require customized solutions; for example, aircraft flight controls. Single-board computers using semi-customized or fully proprietary hardware may be chosen for very demanding control applications where the high development and maintenance cost can be supported. "Soft PLCs" running on desktop-type computers can interface with industrial I/O hardware while executing programs within a version of commercial operating systems adapted for process control needs.[9]

Programmable controllers are widely used in motion control, positioning control and torque control. Some manufacturers produce motion control units to be integrated with PLC so that G-code (involving a CNC machine) can be used to instruct machine movements.

PLCs may include logic for single-variable feedback analog control loop, a "proportional, integral, derivative" or "PID controller". A PID loop could be used to control the temperature of a manufacturing process, for example. Historically PLCs were usually configured with only a few analog control loops; where processes required hundreds or thousands of loops, a distributed control system (DCS) would instead be used. As PLCs have become more powerful, the boundary between DCS and PLC applications has become less distinct.

PLCs have similar functionality as Remote Terminal Units. An RTU, however, usually does not support control algorithms or control loops. As hardware rapidly becomes more powerful and cheaper, RTUs, PLCs and DCSs are increasingly beginning to overlap in responsibilities, and many vendors sell RTUs with PLC-like features and vice versa. The industry has standardized on the IEC 61131-3 functional block language for creating programs to run on RTUs and PLCs, although nearly all vendors also offer proprietary alternatives and associated development environments.

In recent years "Safety" PLCs have started to become popular, either as standalone models (Pilz PNOZ Multi, Sick etc.) or as functionality and safety-rated hardware added to existing controller architectures (Allen Bradley Guardlogix, Siemens F-series etc.). These differ from conventional PLC types as being suitable for use in safety-critical applications for which PLCs have traditionally been supplemented with hard-wired safety relays. For example, a Safety PLC might be used to control access to a robot cell with trapped-key access, or perhaps to manage the shutdown response to an emergency stop on a conveyor production line. Such PLCs typically have a restricted regular instruction set augmented with safety-specific instructions designed to interface with emergency stops, light screens and so forth. The flexibility that such systems offer has resulted in rapid growth of demand for these controllers.

Discrete and analog signals

Discrete signals behave as binary switches, yielding simply an On or Off signal (1 or 0, True or False, respectively). Push buttons, Limit switches, and photoelectric sensors are examples of devices providing a discrete signal. Discrete signals are sent using either voltage or current, where a specific range is designated as *On* and another as *Off*. For example, a PLC might use 24 V DC I/O, with values above 22 V DC representing *On*, values below 2VDC representing *Off*, and intermediate values undefined. Initially, PLCs had only discrete I/O.

Analog signals are like volume controls, with a range of values between zero and full-scale. These are typically interpreted as integer values (counts) by the PLC, with various ranges of accuracy depending on the device and the number of bits available to store the data. As PLCs typically use 16-bit signed binary processors, the integer values are limited between -32,768 and +32,767. Pressure, temperature, flow, and weight are often represented by analog signals. Analog signals can use voltage or current with a magnitude proportional to the value of the process signal. For example, an analog 0 - 10 V input or 4-20 mA would be converted into an integer value of 0 - 32767.

Current inputs are less sensitive to electrical noise (i.e. from welders or electric motor starts) than voltage inputs.

Example

As an example, say a facility needs to store water in a tank. The water is drawn from the tank by another system, as needed, and our example system must manage the water level in the tank by controlling the valve that refills the tank. Shown is a "ladder diagram" which shows the control system. A ladder diagram is a method of drawing control circuits which pre-dates PLCs. The ladder diagram resembles the schematic diagram of a system built with electromechanical relays. Shown are:

- Two inputs (from the low and high level switches) represented by contacts of the float switches
- An output to the fill valve, labelled as the fill valve which it controls
- An "internal" contact, representing the output signal to the fill valve which is created in the program.
- A logical control scheme created by the interconnection of these items in software

In the ladder diagram, the contact symbols represent the state of bits in processor memory, which corresponds to the state of physical inputs to the system. If a discrete input is energized, the memory bit is a 1, and a "normally open" contact controlled by that bit will pass a logic "true" signal on to the next element of the ladder. Internal status bits, corresponding to the state of discrete outputs, are also available to the program.

In the example, the physical state of the float switch contacts must be considered when choosing "normally open" or "normally closed" symbols in the ladder diagram. The PLC has two discrete inputs from float switches (Low Level and High Level). Both float switches close their contacts when the water level in the tank is above the physical location of the switch.

When the water level is below both switches, the float switch physical contacts are open, and neither PLC input is energized. The "normally closed" contact symbol is used for both "Low Level" and "High Level" to pass a logic "true" signal through the rung of the ladder. This sets the output status bit of the PLC to true (logic 1) and then the

PLC uses a discrete output to open the inlet valve to fill the tank. The internal "Fill Valve" contact latches the circuit so that even when the "Low Level" input goes away as the water begins to rise, the fill valve remains on. Once the water level rises enough so that the "High Level" switch is off (up), the PLC will shut the inlet to stop the water from overflowing; This is an example of seal-in (latching) logic. The output is sealed in until a high level condition breaks the circuit. After that the fill valve remains closed (inactive) until the level drops so low that the low level switch is activated, and the described process is repeated.

```
|                                                                   |
|   Low Level      High Level                     Fill Valve        |
|------[/]------|------[/]----------------------(OUT)---------|
|               |                                                   |
|               |                                                   |
|               |                                                   |
|   Fill Valve  |                                                   |
|------[ ]------|                                                   |
|                                                                   |
|                                                                   |
```

A complete program may contain thousands of rungs, evaluated in sequence. Typically the PLC processor will alternately scan all its inputs and update outputs, then evaluate the ladder logic; input changes during a program scan will not be effective until the next I/O update. A complete program scan may take only a few milliseconds, much faster than changes in the controlled process.

Programmable controllers vary in their capabilities for a "rung" of a ladder diagram. Some only allow a single ouput bit. There are typically limits to the number of series contacts in line, and the number of branches that can be used. Each element of the rung is evaluated sequentially. If elements change their state during evaluation of a rung, hard-to-diagnose faults can be generated, although sometimes (as above) the technique is useful. Some implementations forced evaluation from left-to-right as displayed and did not allow reverse flow of a logic signal (in multi-branched rungs) to affect the output.

See also

- Industrial control systems
- Industrial safety systems

References

[1] E. A. Parr, *Industrial Control Handbook*, Industrial Press Inc., 1999 ISBN 0-8311-3085-7

[2] M. A. Laughton, D. J. Warne (ed), *Electrical Engineer's Reference book, 16th edition*,Newnes, 2003 Chapter 16 *Programmable Controller*

[3] "The father of invention: Dick Morley looks back on the 40th anniversary of the PLC" (http://www.automationmag.com/programable-control/features/the-father-of-invention-dick-morley-looks-back-on-the-40th-anniversary-of-the-plc.html). *Manufacturing Automation*. 12 September 2008. .

[4] Harms, Toni M. & Kinner, Russell H. P.E., *Enhancing PLC Performance with Vision Systems*. 18th Annual ESD/HMI International Programmable Controllers Conference Proceedings, 1989, p. 387-399.

[5] Maher, Michael J. *Real-Time Control and Communications*. 18th Annual ESD/SMI International Programmable Controllers Conference Proceedings, 1989, p. 431-436.

[6] Kinner, Russell H., P.E. *Designing Programable Controller Application Programs Using More than One Designer*. 14th Annual International Programmable Controllers Conference Proceedings, 1985, p. 97-110.

[7] W. Bolton, *Programmable Logic Controllers, Fifth Edition*, Newnes, 2009 ISBN 978-1-85617-751-1, Chapter 1

[8] Keller, William L Jr. *Grafcet, A Functional Chart for Sequential Processes*, 14th Annual International Programmable Controllers Conference Proceedings, 1984, p. 71-96.

[9] Gregory K. McMillan, Douglas M. Considine (ed), *Process/Industrial Instruments and Controls Handbook Fifth Edition*, McGraw-Hill, 1999 ISBN 0-07-012582-1 Section 3 *Controllers*

Further reading

- Daniel Kandray, *Programmable Automation Technologies*, Industrial Press, 2010 ISBN 978-0-8311-3346-7, Chapter 8 *Introduction to Programmable Logic Controllers*

External links

- Programmable logic controller (http://www.dmoz.org/Business/Electronics_and_Electrical/Control_Systems/Programmable_Logic_Controllers/) at the Open Directory Project

Northeast_blackout_of_2003

The **Northeast blackout of 2003** was a widespread power outage that occurred throughout parts of the Northeastern and Midwestern United States and Ontario, Canada, on Thursday, August 14, 2003, just before 4:10 p.m. EDT (UTC−04).[1] While some power was restored by 11 p.m., many did not get power back until two days later.[2] At the time, it was the second most widespread blackout in history, after the 1999 Southern Brazil blackout.[3] [4] The blackout affected an estimated 10 million people in Ontario and 45 million people in eight U.S. states.

States and provinces that experienced the blackout; not all areas within these political boundaries were affected.

Immediate impact

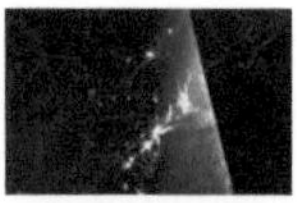

NOAA satellite imagery one day before and the night of the blackout.

According to the New York Independent System Operator (or NYISO) – the ISO responsible for managing the New York state power grid – a 3,500 MW power surge (towards Ontario) affected the transmission grid at 4:10:39 p.m.

EDT.[5] From then through about 4:40 p.m. EDT, outages were reported in Cleveland, Akron, Toledo, New York City, Westchester, Orange and Rockland Counties, Baltimore, Buffalo, Rochester, Syracuse, Binghamton, Albany, Detroit, and parts of New Jersey, including the city of Newark. This was followed by other areas initially unaffected, including all of New York City, portions of southern New York state, New Jersey, Vermont, Connecticut, and most of Ontario, Canada.[6] Eventually a large, somewhat triangular area bounded by Lansing, Michigan, Sault Ste. Marie, Ontario, the shore of James Bay, Ottawa, New York, and Toledo was left without power. According to the official analysis of the blackout prepared by the US and Canadian governments, more than 508 generating units at 265 power plants shut down during the outage. In the minutes before the event, the NYISO-managed power system was carrying 28,700 MW of load. At the height of the outage, the load had dropped to 5,716 MW, a loss of 80%.[5]

Within the area affected, about 200,000 people continued to have power – in the Niagara Peninsula of Ontario; the easternmost corner of Ontario (centred on Cornwall); the portion of New York state including parts of Albany and north and west of Albany; a small pocket of mid-east Michigan; the Upper Peninsula of Michigan, and small pockets in New Jersey. The unaffected area was protected by transmission circuit devices at the Sir Adam Beck Hydroelectric Power Stations in Niagara Falls, at a switching station of the hydroelectric power station in Cornwall, as well as central New York state. Philadelphia and the surrounding mid-Atlantic areas were also completely unaffected because PJM disconnected them.[5]

Essential services remained in operation in some of these areas. In others, backup generation systems failed. Telephone networks generally remained operational, but the increased demand triggered by the blackout left many circuits overloaded. Water systems in several cities lost pressure, forcing boil-water advisories to be put into effect. Cellular service was interrupted as mobile networks were overloaded with the increase in volume of calls. Major cellular providers continued to operate on standby generator power. Television and radio stations remained on the air, with the help of backup generators, although some stations were knocked off the air for periods ranging from several hours to the length of the entire blackout.

It was a hot day (over 31 °C or 88 °F) in much of the affected region, and the heat played a role in the initial event that triggered the wider power outage. The high ambient temperature increased energy demand, as people across the region turned on fans and air conditioning. This caused the power lines to sag as higher currents heated the lines.

In areas where power remained off after nightfall, the Milky Way and orbiting artificial satellites became visible to the naked eye in metropolitan areas where they cannot ordinarily be seen due to the effects of particulate air pollution and light pollution.[7]

Most of the Amtrak Northeast Corridor service was interrupted, as it uses electric locomotives; electrified commuter railways also shut down.[6] Via Rail in Canada was able to continue most of its service. The power outage's effects on international air transport and financial markets were widespread. The reliability of the electrical grid was called into question.

Media coverage and official reports

In the United States and Canada, the regional blackout dominated news broadcasts and news headlines beginning August 15. U.S. and Canadian broadcast media pre-empted normal programming in favor of full-time, advertising-free coverage of the unfolding story. Once terrorism had been conclusively ruled out as a cause, many stations switched back to normal programming following an 8:30 p.m. EDT address by President George W. Bush.[8] National news stations, such as the CBC and CNN, continued to cover the story by inviting politicians and electrical experts to discuss the situation and suggest ways to prevent blackouts. Internationally, coverage of the story focused on the development of the situation in the New York City metropolitan area.

Statements made in the aftermath

During the first two hours of the event, various officials offered speculative explanations as to its root cause:

- Official reports from the office of Canadian Prime Minister Jean Chrétien stated that lightning had struck a power plant in northern New York, resulting in a cascading failure of the surrounding power grid and wide-area electric power transmission grid. However, power officials in the State of New York responded by stating that the problem did not originate in the United States, that there was no rain storm in the area where the lightning strike was supposed to have taken place, and that the power plant in question remained in operation throughout the blackout.
- Canadian Defence Minister John McCallum blamed an outage at a nuclear plant in Pennsylvania, but that state's authorities reported that all the plants were functioning normally. McCallum later stated that his sources had given him incorrect information.
- New York state Governor George Pataki blamed the power outage on Canada, stating "the New York independent systems operator says they are virtually certain it had nothing to do in New York state. And they believe it occurred west of Ontario, cascaded from there into Ontario, Canada, and through the northeast."[9] This was later proven to be false.
- CNN cited unnamed officials as saying that the Niagara-Mohawk power grid, which provides power for large portions of New York and parts of Canada, was overloaded. Between 4:10 and 4:13 p.m. EDT, 21 power stations throughout that grid shut down.[6]
- New Mexico governor Bill Richardson, who formerly headed the Department of Energy, in a live television interview 2 hours into the blackout characterized the United States as "a superpower with a third-world electricity grid." In Europe, this statement was published accompanied with comparisons highlighting the tighter, safer and better interconnected European electricity network (though it would suffer a similar blackout six weeks later).
- In the ensuing days, critics focused on the role of electricity market deregulation for the inadequate state of the electric power transmission grid, claiming that deregulation laws and electricity market mechanisms have failed to provide market participants with sufficient incentives to construct new transmission lines and maintain system security.
- Later that night, claims surfaced that the blackout may have started in Ohio up to one hour before the network shut down, a claim denied by Ohio's FirstEnergy utility.
- The president of the North American Electric Reliability Corporation said that the problem originated in Ohio.[10]
- By Saturday morning, investigators believed that the problem began with a sudden shift in the direction of power flow on the northern portion of the Lake Erie Transmission Loop, a system of transmission lines that circles Lake Erie on both U.S. and Canadian soil.
- Per Christopher Mark Heben, US Navy Seal, on a National Public Radio *How To Do Everything* Podcast, the blackout occurred because a military intelligence aircraft pilot "hit the wrong switch as they flew from North Carolina to Maine and basically zapped the whole power grid down with an electronic counter measure device."[11]

Causes

Background

Electrical power cannot easily be stored over extended periods of time, and is generally consumed less than a second after being produced. The load on any network must be matched by the supply to it and its ability to transmit that power. Any overload of a power line, or underload/overload of a generator, can cause hard-to-repair and costly damage, so the affected device is disconnected from the network if a serious imbalance is detected.

As power lines carry more current, they get hotter. This causes the metal conductors in the lines to lengthen and sag between towers. They may safely reach a specified minimum clearance height above the ground. If the lines sag lower, a flashover to nearby objects (such as trees) can occur, causing a transient increase in current. Automatic protective relays detect the high current and quickly act to disconnect the faulted line from service. To maintain the lines' specified operating clearance, the right-of-way must be kept clear of vegetation.

Should a fault occur and take a line out of service, the change in current flow is compensated by other transmission lines, which must have enough spare capacity to carry the excess current. If they do not, overload protection in those lines will also trip, causing a cascading failure as the excess current is switched onto neighbouring circuits running at or near their capacity.

System operators are responsible for ensuring that power supply and loads remain balanced, and for keeping the system within safe operational limits such that no single fault can cause the system to fail. After a failure affecting their system, operators must obtain more power from generators or other regions or "shed load" (meaning to intentionally cut power to some areas, or reducing voltage to a given area, creating a brownout) until they can be sure that the worst remaining possible failure anywhere in the system will not cause a system collapse. In an emergency, they are expected to immediately shed load as required to bring the system into balance.

To assist the operators there are computer systems, with backups, which issue alarms when there are faults in the transmission or generation system. Power flow modeling tools let them analyze the current state of their network, predict whether any parts of it may be overloaded, and predict what the worst possible failure left is, so that they can change the distribution of generation or reconfigure the transmission system to prevent a failure should this situation occur. If the computer systems and their backups fail, the operators are required to monitor the grid manually, instead of relying on computer alerts. If they cannot interpret the current state of the power grid in such an event, they follow a contingency plan, contacting other plant and grid operators by telephone if necessary. If there is a failure, they are also required to notify adjacent areas which may be affected, so those can predict the possible effects on their own systems.

Local operators are co-ordinated by regional centers, but the operating principle is the same whether the network is large or small.

Investigation efforts

A joint federal task force was formed by the governments of Canada and the U.S. to oversee the investigation and report directly to Ottawa and Washington. The task force was led by then-Canadian Natural Resource Minister Herb Dhaliwal and U.S. Energy Secretary Spencer Abraham.

In addition to determining the initial cause of the cascading failure, the investigation of the incident also included an examination of the failure of safeguards designed to prevent a repetition of the Northeast blackout of 1965. Issues of failure to maintain the electrical infrastructure, failure of upgrading to so-called "smart cables," failure of shunting and rerouting mechanisms, AC vs. DC intersystem ties, and substitution of electricity market forces for central planning were expected to arise. The North American Electric Reliability Corporation, a joint Canada-U.S. council, is responsible for dealing with these issues.

On November 19, 2003, U.S. Energy Secretary Spencer Abraham said his department would not seek to punish FirstEnergy Corp for its role in the blackout because current U.S. law does not require electric reliability standards. Abraham stated, "The absence of enforceable reliability standards creates a situation in which there are limits in terms of federal level punishment."[12]

Findings

In February 2004, the U.S.-Canada Power System Outage Task Force released their final report, placing the causes of the blackout into four groups:[13]

First, that FirstEnergy (FE) and its reliability council "failed to assess and understand the inadequacies of FE's system, particularly with respect to voltage instability and the vulnerability of the Cleveland-Akron area, and FE did not operate its system with appropriate voltage criteria".

Second, that FirstEnergy "did not recognize or understand the deteriorating condition of its system".

Third, that FirstEnergy "failed to manage adequately tree growth in its transmission rights-of-way".

Finally, the "failure of the interconnected grid's reliability organizations to provide effective real-time diagnostic support."

The report states that a generating plant in Eastlake, Ohio (a suburb of Cleveland) went offline amid high electrical demand, putting a strain on high-voltage power lines (located in a distant rural setting) which later went out of service when they came in contact with "overgrown trees". The cascading effect that resulted ultimately forced the shutdown of more than 100 power plants.

This trip caused load to transfer to other transmission lines, which were not able to bear the load, tripping their relays. Once these multiple trips occurred, multiple generators suddenly lost parts of their loads, so they accelerated out of phase with the grid at different rates, and tripped out to prevent damage.

Computer failure

A software bug known as a race condition existed in General Electric Energy's Unix-based XA/21 energy management system. Once triggered, the bug stalled FirstEnergy's control room alarm system for over an hour. System operators were unaware of the malfunction; the failure deprived them of both audio and visual alerts for important changes in system state.[14] [15] After the alarm system failure, unprocessed events queued up and the primary server failed within 30 minutes. Then all applications (including the stalled alarm system) were automatically transferred to the backup server, which itself failed at 14:54. The server failures slowed the screen refresh rate of the operators' computer consoles from 1–3 seconds to 59 seconds per screen. The lack of alarms led operators to dismiss a call from American Electric Power about the tripping and reclosure of a 345 kV shared line in northeast Ohio. Technical support informed control room personnel of the alarm system failure at 15:42.[16]

Sequence of events

The following is the blackout's sequence of events on August 14, 2003[17] [18] [19] (times in EDT):

- 12:15 p.m. Incorrect telemetry data renders inoperative the state estimator, a power flow monitoring tool operated by the Indiana-based Midwest Independent Transmission System Operator (MISO). An operator corrects the telemetry problem but forgets to restart the monitoring tool.
- 1:31 p.m. The Eastlake, Ohio generating plant shuts down. The plant is owned by FirstEnergy, an Akron, Ohio-based company that had experienced extensive recent maintenance problems.
- 2:02 p.m. The first of several 345 kV overhead transmission lines in northeast Ohio fails due to contact with a tree in Walton Hills, Ohio.[20] 41°21′22″N 81°34′10″W
- 2:14 p.m. An alarm system fails at FirstEnergy's control room and is not repaired.

- 3:05 p.m. A 345 kV transmission line known as the Chamberlin-Harding line sags into a tree and trips in Parma, south of Cleveland.
- 3:17 p.m. Voltage dips temporarily on the Ohio portion of the grid. Controllers take no action.
- 3:32 p.m. Power shifted by the first failure onto another 345 kV power line, the Hanna-Juniper interconnection, causes it to sag into a tree, bringing it offline as well. While MISO and FirstEnergy controllers concentrate on understanding the failures, they fail to inform system controllers in nearby states.
- 3:39 p.m. A FirstEnergy 138 kV line trips in northern Ohio.[21]
- 3:41 p.m. A circuit breaker connecting FirstEnergy's grid with that of American Electric Power is tripped as a 345 kV power line (Star-South Canton interconnection) and fifteen 138 kV lines fail in rapid succession in northern Ohio.
- 3:46 p.m. A fifth 345 kV line, the Tidd-Canton Central line, trips offline.
- 4:05:57 p.m. The Sammis-Star 345 kV line trips due to undervoltage and overcurrent interpreted as a short circuit. Later analysis suggests that the blackout could have been averted prior to this failure by cutting 1.5 GW of load in the Cleveland–Akron area.
- 4:06–4:08 p.m. A sustained power surge north toward Cleveland overloads three 138 kV lines.
- 4:09:02 p.m. Voltage sags deeply as Ohio draws 2 GW of power from Michigan, creating simultaneous undervoltage and overcurrent conditions as power attempts to flow in such a way as to rebalance the system's voltage.
- 4:10:34 p.m. Many transmission lines trip out, first in Michigan and then in Ohio, blocking the eastward flow of power around the south shore of Lake Erie from Toledo, Ohio, east through Erie, Pennsylvania and into the Buffalo, New York metropolitan area. Suddenly bereft of demand, generating stations go offline, creating a huge power deficit. In seconds, power surges in from the east, overloading east-coast power plants whose generators go offline as a protective measure, and the blackout is on.
- 4:10:37 p.m. The eastern and western Michigan power grids disconnect from each other. Two 345 kV lines in Michigan trip. A line that runs from Grand Ledge to Ann Arbor known as the Oneida-Majestic interconnection trips. A short time later, a line running from Bay City south to Flint in Consumers Energy's system known as the Hampton-Thetford line also trips.
- 4:10:38 p.m. Cleveland separates from the Pennsylvania grid.
- 4:10:39 p.m. 3.7 GW power flows from the east along the north shore of Lake Erie, through Ontario to southern Michigan and northern Ohio, a flow more than ten times greater than the condition 30 seconds earlier, causing a voltage drop across the system.
- 4:10:40 p.m. Flow flips to 2 GW eastward from Michigan through Ontario (a net reversal of 5.7 GW of power), then reverses back westward again within a half second.
- 4:10:43 p.m. International connections between the United States and Canada start to fail.
- 4:10:45 p.m. Northwestern Ontario separates from the east when the Wawa-Marathon 230 kV line north of Lake Superior disconnects. The first Ontario power plants go offline in response to the unstable voltage and current demand on the system.
- 4:10:46 p.m. New York separates from the New England grid.
- 4:10:50 p.m. Ontario separates from the western New York grid.
- 4:11:57 p.m. The Keith-Waterman, Bunce Creek-Scott 230 kV lines and the St. Clair–Lambton #1 230 kV line and #2 345 kV line between Michigan and Ontario fail.
- 4:12:03 p.m. Windsor, Ontario and surrounding areas drop off the grid.
- 4:12:58 p.m. Northern New Jersey separates its power-grids from New York and the Philadelphia area, causing a cascade of failing secondary generator plants along the New Jersey coast and throughout the inland regions west.
- 4:13 p.m. End of cascading failure. 256 power plants are off-line, 85% of which went offline after the grid separations occurred, most due to the action of automatic protective controls.

Effects

Major cities affected

City	Number of people affected
New York City and surrounding areas	14,300,000
Greater Toronto Area (Golden Horseshoe)	8,300,000
Newark, New Jersey and surrounding counties and suburbs	6,980,000
Detroit and Surrounding Areas	5,400,000
Cleveland and Greater Cleveland	2,900,000
Ottawa	780,000 of 1,120,000*
Buffalo and Surrounding Areas	1,100,000
Rochester	1,050,000
Baltimore and Surrounding Counties	710,000
London, Ont. and Surrounding Areas	475,000
Toledo	310,000
Windsor	208,000
Estimated Total[22]	55,000,000
*Ottawa-Gatineau is a special case in that it is divided by a provincial boundary and the Ontario and Quebec grids are not synchronously connected. Gatineau had power. One may have seen the drastic cutoff of areas still having power when they were crossing the Portage Bridge between Gatineau and Ottawa - the cutoff was at the provincial line (street lights on the bridge were still lit on the Quebec side of the structure.)	

Affected infrastructure

A streetcar left stranded by the blackout in Toronto

Power generation

With the power fluctuations on the grid, power plants automatically went into "safe mode" to prevent damage in the case of an overload. This put much of the nuclear power normally available offline until those plants could be slowly taken out of "safe mode". In the meantime, all available hydro-electric plants (as well as many coal and oil fired plants) were brought online, bringing some electrical power to the areas immediately surrounding the plants by the morning of August 15. Homes and businesses both in the affected area and in nearby areas were requested to limit power usage until the grid was back to full power.

Water supply

Some areas lost water pressure because pumps lacked power. This loss of pressure caused potential contamination of the water supply. Four million customers of the Detroit water system in eight counties were under a boil water advisory until 18 August, four days after the initial outage. One county, Macomb, ordered all 2,300 restaurants closed until they were decontaminated after the advisory was lifted. Twenty people living on the St. Clair River claim to have been sickened after bathing in the river during the blackout. The accidental release of 140 kg (310 lb) of vinyl chloride from a Sarnia, Ontario chemical plant was not revealed until five days later. Cleveland also lost water pressure and instituted a boil water advisory. Cleveland and New York had sewage spills into waterways, requiring beach closures. Newark, New Jersey and northern cities had major sewage spills into the Passaic and Hackensack Rivers, which flow directly to the Atlantic Ocean. Kingston lost power to sewage pumps, causing raw waste to be dumped into the Cataraqui River at the base of the Rideau Canal.

Transportation

Amtrak's Northeast Corridor railroad service was stopped north of Philadelphia, Pennsylvania, and all trains running into and out of New York City were shut down, initially including the Long Island Rail Road and the Metro-North Railroad; both were able to establish a bare-bones "all-diesel" service by the next morning. Canada's Via Rail, which services Toronto and Montreal, suffered service delays, but most routes were still running, and normal service was resumed on most Via routes by the next morning.

Passenger screenings at affected airports ceased. Regional airports were shut down for this reason. In New York, flights were cancelled even after power had been restored to the airports because of difficulties accessing "electronic-ticket" information. Air Canada flights remained grounded on the morning of August 15 due to reliable power not having been restored to its Mississauga control center. It expected to resume operations by midday. This problem affected all Air Canada service and canceled the most heavily traveled flights to Halifax and Vancouver. At Chicago's Midway International Airport, Southwest Airlines employees spent 48 hours dealing with the disorder caused by the blackout's sudden incidence.[23] Many gas stations were unable to pump fuel due to lack of electricity. In North Bay, Ontario, for instance, a long line of transport trucks was held up, unable to go further west to Manitoba without refueling. In some cities, traffic problems were compounded by motorists who simply drove until their cars ran out of gas on the highway. Gas stations operating in pockets of Burlington, Ontario, that had power were reported to be charging prices up to 99.9 cents/litre ($3.776 per US gallon) when the going rate prior to the blackout was lower than 70 cents/litre. Customers still lined up for hours to pay prices most people considered price gouging. Station operators claimed that they had a limited supply of gasoline and did not know when their tanks would be refilled, prompting the drastic price increases.

Many oil refineries on the East Coast of the United States shut down as a result of the blackout, and were slow to resume gasoline production. As a result, gasoline prices were expected to rise approximately 10 cents/gallon (3 c/L) in the United States. In Canada, gasoline rationing was also considered by the authorities.

Communication

Cellular communication devices were disrupted. This was mainly due to the loss of backup power at the cellular sites where generators ran out of fuel or cell phone batteries ran out of charge. Wired telephone lines continued to work, although some systems were overwhelmed by the volume of traffic, and millions of home users had only cordless telephones depending on house current. Most New York and many Ontario radio stations were momentarily knocked off the air but were able to return with backup power.

Cable television systems were disabled, and areas that had power restored (and had power to their television sets) could not receive information until power was restored to the cable provider. Those who relied on the Internet were similarly disconnected from their news source for the duration of the blackout, with the exception of dialup access from laptop computers, which were widely reported to work until the batteries ran out of charge.

The blackout impacted communications well outside the immediate area of power outage. The New Jersey-based internet operations of Advance Publications were among those knocked out by the blackout. As a result, the internet editions of Advance newspapers as far removed from the blackout area as *The Birmingham News*, the *New Orleans Times Picayune*, and *The Oregonian* were offline for days.[24]

Amateur radio operators passed emergency communications during the blackout.[25]

Industry

Large numbers of factories were closed in the affected area and others outside the area were forced to close or slow work because of supply problems and the need to conserve energy while the grid was being stabilized. At one point a 7-hour wait developed for trucks crossing the Ambassador Bridge between Detroit and Windsor due to the lack of electronic border check systems. Freeway congestion in affected areas affected the "just in time" (JIT) supply system. Some industries (including the auto industry) did not return to full production until 22 August.

By region

New York, United States

People walking in New York City during the blackout

Almost the entire state of New York lost power. Exceptions include Freeport and Rockville Centre on Long Island, which relied on localized power plants; the Capital District, where power dipped briefly but returned, (a few areas of which, such as portions of Latham, NY, never lost service during the entire event); the southernmost areas of the Southern Tier of Upstate New York, mostly near Waverly, which relied on power from Pennsylvania; the city of Plattsburgh; Starrett City, Brooklyn, which has auxiliary power; most of the city of Buffalo; and pockets of Amherst in the Buffalo area, running off university power. There were also some small pockets of power in the suburbs of Rochester, as a few smaller power companies operating in those areas were able to keep running. The North Shore Towers complex was unaffected by the blackout due to their on-site self-generating power plant. Power was also available at the Kodak Park facility and its surrounding neighborhood in the city. Power was lost at the Oak Hill Country Club, in nearby Pittsford, New York, where the PGA Championship was being played, which caused minor interruptions to the tournament. Also, that evening's Major League Baseball game between the New York Mets and the San Francisco Giants at Shea Stadium was postponed. In New York, all prisons were blacked out and switched to generator power. The two Indian Point nuclear reactors on the Hudson River near Peekskill, the two reactors at Nine Mile Point nuclear plant, the single reactor at Ginna nuclear plant near Rochester, and the FitzPatrick reactor near Oswego all shut down. With three other nuclear plants shut down in Ohio, Michigan, and New Jersey, a total of nine reactors were affected. The governor of New York state, George Pataki, declared a state of emergency.

Manhattan, including Wall Street and the United Nations, was completely shut down, as were all area airports (with the exception of Newburgh's Stewart Airport, which had once been a military airbase and had its own generators), and all New York area rail transportation including the subway, the PATH lines between Manhattan and New Jersey, New Jersey Transit's Northeast Corridor and North Jersey Coast lines between New York Penn Station and Newark Penn Station, the Morris & Essex Lines between New York Penn Station and Newark Broad Street, the Metro-North Railroad, and the Long Island Rail Road. Hundreds of people were trapped in elevators; by late evening the New York City Fire Department had reportedly confirmed that all stalled elevators in approximately 800 Manhattan high-rise office and apartment buildings had been cleared. More than 600 subway and commuter rail cars were trapped between stations; the NY State Metropolitan Transportation Authority and the Port Authority of New York and New Jersey—which operates the PATH lines—reported that all passengers were evacuated without serious injury. However, PATH was first to resume subway service on Sixth Avenue (albeit on 15-minute headways) by 6PM that evening.

Without traffic lights, gridlock was reported as people in lower and midtown Manhattan fled their offices on foot; for hours into the evening the streets, highways, bridges and tunnels were jammed with traffic and pedestrians leaving Manhattan, though many civilians opted to help direct traffic. The bus journey from Manhattan to Washington, D.C. (which normally takes four hours) took more than eight hours, with reports that it took four hours just to get out of Manhattan. Mayor Michael Bloomberg advised residents to open their windows, drink plenty of liquids to avoid heat stroke in the heat, and not to forget their pets. Temperatures were 92 °F (33 °C) with high humidity, as New York had just experienced a record-breaking rain spell that had started at the end of July. With cell phone operation mostly stalled by circuit overloads, New Yorkers were lining up 10 deep or more at pay phones as ordinary telephone service remained largely unaffected.

While some commuters were able to find alternate sleeping arrangements, many were left stranded in New York and slept in parks and on the steps of public buildings. While practically all businesses and retail establishments closed down, many bars and pubs reported a brisk business as many New Yorkers took the opportunity to spend the evening "enjoying" the blackout. Since most perishable items were going to spoil anyway, many restaurants and citizens simply prepared what they could and served it to anyone who wanted it, leading to vast block parties in many New York neighborhoods. Any ice cream in frozen storage also had to be quickly served to any and all passers by. The Indigo Girls were scheduled to perform that evening at Central Park SummerStage, and the band took the stage as planned to play one of the only shows in the affected area, using generators that had been filled with fuel that morning. Their performance was not a full-length concert, as they needed to end early to allow all attendees to exit Central Park before dark. The venue also had bathrooms and vendors cooking food on propane grills.

40,000 police and the entire fire department were called in to maintain order. At least two fatalities were linked to the use of flames to provide light, and many nonfatal fires also resulted from the use of candles. New York City's Office of Emergency Management activated its Emergency Operations Center, from which more than 70 agencies coordinated response efforts, which included delivery of portable light towers to unlit intersections, generators and diesel fuel to hospitals, and a portable steam generator necessary to power air conditioning units at the American Stock Exchange.

Verizon's emergency generators failed several times, leaving the emergency services number 9-1-1 out of service for several periods of about a quarter hour each. New York City's 311 information hotline received over 175,000 calls from concerned residents during the weekend. Amateur radio operators attached to New York City ARES provided a backup communications link to emergency shelters and hospitals. Amateur radio repeaters were supplied with emergency power via generators and batteries and remained functional.

Many major U.S. networks (CBS, NBC, ABC, and FOX) and some cable TV networks (such as HBO, MTV, and Nickelodeon) were unable to broadcast because of the lack of electricity in the New York City area, but back-up stations in Dallas and flagship transmitters there made it possible for prime-time television to be broadcast. (ABC chose instead to cover the news from Washington, D.C. during the blackout).

For delayed effects at Niagara Falls, see below under Ontario.

New Jersey, United States

Affected areas included most of Hudson, Morris, Essex, Union, Passaic and Bergen Counties, including the major cities of Paterson and Newark, although some sections of Newark and East Orange still had power; also, small sections of certain towns in Essex and Hudson Counties had power. Power was returned first to the urban areas because of concerns of safety and unrest. Counties as far south as Cumberland were affected, but power was restored within an hour. Some towns in Bergen County saw only a momentary dropout in power, but did see wild oscillations in powerline voltage, ranging from about 90V to 135V, up and down every few minutes for an hour.

The day following the blackout, August 15, the New Jersey Turnpike stopped collecting tolls until 9:00 a.m.

Baltimore, United States

The outage affected many households and businesses in Baltimore City and all of the surrounding counties including Baltimore County. Flights were canceled and rerouted at the airports in the Baltimore-Washington area, including Baltimore-Washington International Thurgood Marshall Airport (BWI), Washington Dulles International Airport, and Ronald Reagan Washington National Airport.

Connecticut, United States

Parts of New London, Hartford, New Haven, Litchfield and Fairfield Counties, from Greenwich to Danbury and Bridgeport, were affected, although most of the state had power all evening, aside from a few momentary interruptions that caused computers to reboot. Metro North trains stopped, and remained on the tracks for hours, until they could be towed to the nearest station. Generally, most of the state east of Interstate 91, and some places west of I-91, had power during the duration of the blackout, with some of New Haven's eastern suburbs being seen as the easternmost extreme of the effects of the blackout.

A local controversy ensued in the days after the blackout, when the federal government ordered power companies to energize the HVDC Cross Sound Cable between New Haven and Long Island. This cable had been installed, but had not been activated due to environmental and fisheries concerns. The Attorney General of Connecticut, Richard Blumenthal, and the Governor of New York, George Pataki, traded insults over the cable. Connecticut politicians expressed their outrage that the cable was being turned on, since it did not help anyone in Connecticut, as the cable would transport power from Connecticut to Long Island.

Massachusetts, United States

A small area of extreme western Massachusetts was affected. In Worcester the event was of sufficient magnitude to reboot some computers, while in Springfield the effect of the event was enough to cause the automatic startup of commercial and industrial backup generation facilities. Some areas were subjected to lower-than-normal voltage (as low as 100 volts AC) and brownouts for periods of up to 24 hours. The Boston area was spared from the Blackout.

Michigan, United States

About 2.3 million households and businesses were affected, including almost all of Metro Detroit, as well as Lansing, Ann Arbor, and surrounding communities in southeast Michigan. The blackout affected three Michigan utilities; Detroit Edison (whose entire system went down), Lansing Board of Water and Light, and a small portion of Consumers Energy's system in the southeastern corner of the state. Word quickly spread to the surrounding areas without power and many flocked to surrounding areas that still had power, resulting in crowded stores, packed restaurants, booked hotels, and long queues for the gas stations in these towns. Locales closest to the affected areas in the northern Detroit suburbs that did not lose power included the areas of Oxford and Holly, communities along M-24 and M-15, and into the Lapeer and Flint/Tri-Cities area. The city limits of Brighton and Howell were unaffected as well as they received electricity from Consumers Energy via the Genoa-Latson 138KV line which interconnects Detroit Edison and Consumers Energy. Television and radio stations were temporarily knocked off the air and water supplies were disrupted in Detroit due to the failure of electric pumps. Because of the loss of water pressure all water was required to be boiled before use until August 18. Several schools which had planned to begin the school year 18 August were closed until clean water was available. A Marathon Oil refinery in Melvindale, near Detroit, suffered a small explosion from gas buildup, necessitating an evacuation within one mile (1.6 km) around the plant and the closure of Interstate 75. Officials feared the release of toxic gases. Heavy rains on Friday coupled with the lack of sewage pumps closed other expressways and prompted urban flood warnings. Untreated sewage flowed into local rivers in Lansing and Metropolitan Detroit as contingency solutions at some sewage treatment plants failed. In the midst of a summer heat wave, Michiganders were deprived of air conditioning. Several people, mostly elderly individuals, had to be treated for symptoms of heat stroke. In the Detroit area, local television stations' news helicopters were told by each station's management to "stay above the cars' headlights" at night, and to not venture into Downtown Detroit (due to the hazard of flying into an unlit skyscraper). During the days immediately after the blackout, many stations were back on the air, but with limited resources (in one case, WXYZ-TV's news anchor was wearing a T-shirt and shorts, as opposed to his normal news suit, and apologized to viewers for the "rather warm conditions" in the station, as they only had one air conditioner and a couple fans working). The Downriver Communities would also have to contend with basements flooded with sewage-laden water on the weekend immediately after the blackout, with water and sewage pumps off from a lack of power, much to the general annoyance of residents in the areas. News crews of the areas broadcast notices during their coverages of the blackouts to the Downriver residents, explaining why the pumps have shorted out, as well as to limit water usage ("most places have water pressure, some have low pressure...some have none, and some even have *negative pressure*. That means in the next few hours, people in the downriver communities should expect flooded basements from, so move all your valuables high up and out of the basements", as WDIV-TV warned). West Michigan, including the communities of Grand Rapids, Muskegon, and Holland, was mostly unaffected.

Ohio, United States

Over 540,000 homes and businesses were without power. In Cleveland, water service stopped because the city is supplied by electric pumps and backup electricity was available only on a very limited basis and water had to be boiled for several days afterwards. Portions of the cities of Akron, Mansfield, Massillon, Marion and Ashland were without power. Cleveland declared a curfew on all persons under the age of 18. At Cedar Point amusement park in Sandusky, park employees had to help guests walk down the steps of the 205-foot (60 m)-tall Magnum XL-200 roller coaster, which had stopped on the lift hill due to the blackout. Several other guests had to be helped off rides as a result of the blackout. In Toledo, the Mud Hens baseball team postponed the game scheduled for that night. However, some parts of the city were unaffected by the blackout, notably the suburb of Sylvania. Other surrounding cities like Bowling Green only experienced a brief outage.

Ontario, Canada

The area affected by the blackout included all of Southern Ontario, except Niagara Falls and Fort Erie, from Windsor to Ottawa and all the way to the Quebec border. Also affected was Northern Ontario, as far north as Attawapiskat and Moosonee on James Bay and west to Marathon on the Lake Superior shoreline. Communities affected in northern Ontario included Timmins, Cochrane, Sudbury, Kirkland Lake, North Bay, Wawa, and Sault Ste. Marie. Most of Northwestern Ontario (including Thunder Bay) was not affected.

Toronto, Ontario, on the evening of August 14

Traffic lights, the Toronto subway and RT, Toronto streetcar system, the Toronto Stock Exchange, and CBC's studios were shut down in Toronto. The CBC switched to its backup studios in both Calgary and Vancouver for coverage, because newsgathering in Toronto was extremely difficult. The Toronto studios were only run by UPS systems. Many passengers had to be evacuated from subway trains by walking through the tunnels. Major Toronto hospitals reported that they had switched to generators and did not experience problems. The 9-1-1 system was operational. Highway 407, the world's first all-electronic toll highway, was gridlocked with passengers hoping to get a free ride. Parliament Hill was evacuated in Ottawa.

Toronto officials asked residents to curtail unnecessary use of water, as pumps were not working and there was only a 24-hour supply.

Traffic lights, which had no backup power, were all knocked out. All intersections were to be considered an all-way stop. Coupled with the beginning of the evening rush hour, this caused traffic problems. In many major and minor intersections in both large and small cities, such as Ottawa, Toronto, and Burlington, ordinary citizens began directing traffic until police or others relieved them. Since there were not enough police officers to direct traffic at every intersection during the afternoon rush hour, passing police officers distributed fluorescent jackets to people who were directing traffic. Drivers and pedestrians generally followed the instructions from them even though they were not police officers.[3]

Volunteers received fluorescent jackets from the police to direct traffic in Toronto, Ontario during the blackout

Fierce disruptions of truck traffic in northeastern Ontario were reported due to the unavailability of fuel, including the backlog near North Bay. The tunnel and bridge between Windsor and Detroit were also closed, with the bridge's pillars illuminated by emergency floodlights, as to not pose a shipping and airplane hazard.

About 140 miners were marooned underground in the Falconbridge mine in Sudbury when the power went out. Mine officials said that they were safe and could be evacuated if necessary, but were not being evacuated due to the risks of doing so with no power. They were safely evacuated by the morning. In Sarnia, a refinery scrubber lost power and released above-normal levels of pollution; residents were asked to close their windows.

Toronto Union Station during the blackout

On the evening of August 14, Ontario premier Ernie Eves advised Lieutenant Governor James Bartleman to declare a state of emergency, instructing nonessential personnel not to go to work on Friday, August 15. Residents were asked not to use televisions, washing machines, or air conditioners if possible, and warned that some restored power might go off again. Although the full state of emergency was lifted the next day (a Saturday), residents were warned that the normal amount of power would not be available for days, and were still asked to reduce power consumption. The roof of the Skydome in Toronto remained open in an effort to conserve power until August 21, when a thunderstorm struck.

The Toronto Transit Commission operated its streetcars on the Friday, but not on the weekend, and did not reactivate the subway and RT until Monday, August 18, after assurances were received that they would be exempted from any rotating blackouts that might be needed.

Major events such as concerts were canceled for several days. English band Radiohead rescheduled a planned August 16 concert in Toronto until the following October 15 due to the power outage. The opening of the Canadian National Exhibition, scheduled for August 15, was postponed to Tuesday, August 19.

For two days of this recovery period, diversion of water from the Niagara River for hydroelectric generation was increased to the maximum level, normally used only at night and in winter in order to maintain the scenic appearance of Niagara Falls. The resultant drop in the river level below the falls meant that the Maid of the Mist tour boats could not dock safely, and their operation had to be suspended.

The Petro-Canada refinery in Oakville had to perform an emergency shutdown due to the lack of power. The plant's flare system produced large flames during the shutdown, leading to erroneous reports in the media that there had been a fire in the plant. The Petro Canada lubricants plant in Mississauga experienced a fire one week later while restarting normal operations.[26] [27]

Emergency services

In New York, about 3,000 fire calls were reported, many from people using candles. Emergency services responded to 80,000 calls for help, more than double the average.

From 4 PM ET of the 14th to midnight of the 15th, there were 60 all-hands or greater alarm fires; caused mostly by candles. The FDNY answered over 7,500 calls which resulted in the transmission of over 4,000 alarms. www.FDNewYork.com/lightsout.asp [28]

Fatalities

The blackout contributed to at least eleven fatalities,

- In Ontario, three fatalities occurred, a cyclist hit by a car in Guelph[29] and, west of Ottawa, a father of 2 hit by a car in his front yard, and a fire victim. However, the blackout cannot be solely identified as the reason.[30]
- In Connecticut, one fatality was reported.
- In New York City, six fatalities were reported. Two were deaths from carbon monoxide, two from fire, one as a result of a fall from a roof while breaking into a shoe store, and one man died of a heart attack in his neighbor's apartment after climbing the 17 flights to their floor.
- In the Detroit suburb Harper Woods, Michigan, one man was reported dead on WXYZ-TV's news from carbon monoxide poisoning from using a generator inside his house.
- In Pittsfield Township, Michigan, a 27-year-old Belleville, Michigan man was reported unresponsive when a fatal fire destroyed the mobile home that he was sleeping in, according to the *Ann Arbor News*. Officials said the fire was apparently caused when candles were left burning during the blackout.

Long-term effects

The Ontario government fell in a provincial election held in October 2003; consequently, power had long been a major issue. The government may have been hurt by the success of Quebec and Manitoba, which were not affected whereas Ontario was shut down. The extra publicity given to Ontario's need to import electricity from the United States, mostly due to a decision of the government not to expand the province's power generating capabilities, may also have adversely affected the Conservative government. Premier Ernie Eves' handling of the crisis was also criticized; he was not heard from until long after Mayor Bloomberg and Governor Pataki had spoken out. Due to the regular announcements he gave in the days following the blackout, Eves enjoyed a moderate increase in the polls that his party took as a sign of an opportunity to call an election they could win, however that did not prove to be the case.

In the United States, the Bush administration had emphasized the need for changes to the U.S. national energy policy, Critical Infrastructure Protection, and Homeland Security. During the blackout, most systems that would detect unauthorized border crossings, port landings, or detect unauthorized access to many vulnerable sites, failed. There was considerable fear that future blackouts would be exploited for terrorism. In addition, the failure highlighted the ease with which the power grid could be taken down.

Restoration of service

By evening of August 14, power had been restored to:

- Many areas of the Niagara Region in Ontario;
- Areas of the Ontario Golden Horseshoe from St. Catharines to Burlington (supplied from the City of Niagara Falls Ontario, which never lost power);
- Parts of Southwestern Ontario, particularly areas near the Bruce Nuclear Power Plant, lost power for only 4–8 hours;
- parts of Mississauga;
- parts of London, Ontario;
- portions of western Ottawa including Kanata and south to Kingston;
- portions of downtown Toronto;
- Cornwall and Pembroke, Ontario;
- three-quarters of the millions of customers who had lost power in New Jersey;
- parts of Pennsylvania, Ohio and Michigan;
- parts of Long Island;
- Albany and its surroundings;
- New London County, Connecticut;
- Parry Sound, Ontario

Con Edison retracted its claim that New York City would have power by 1 a.m. That night some areas of Manhattan regained power at approximately 5 a.m. (August 15), the New York City borough of Staten Island regained power around 3 a.m. on August 15, and Niagara Mohawk predicted that the Niagara Falls area would have to wait until 8 a.m.

By early evening of August 15, two airports, Cleveland Hopkins International Airport, and Toronto Pearson International Airport were back in service.

Half of the affected part of Ontario had power by the morning of 15 August, though even in areas where it had come back online, some services were still disrupted or running at lower levels. The last areas to regain power were usually suffering from trouble at local electrical substations that was not directly related to the blackout itself.

By 16 August, power was fully restored in New York and Toronto. However, Toronto's subway and streetcars remained out of service until 18 August to prevent the possibility of equipment being stuck in awkward locations if the power was interrupted again. Power had been mostly restored in Ottawa, though authorities warned of possible additional disruptions and advised conservation as power continued to be restored to other areas. Ontarians were asked to reduce their electricity use by 50% until all generating stations could be brought back on line. Four remained out of service on the 19th. Illuminated billboards were largely dormant for the week following the blackout, and many stores had only a portion of their lights on.

Preparations against the possible disruptions threatened by the Year 2000 problem have been credited for the installation of new electrical equipment and systems which allowed for a relatively rapid restoration of power in some areas.

Sustained efforts

In Ontario, some cities took part in power conservation challenges or events to remind citizens of the blackout. The most well known event being the Voluntary Blackout Day hosted by the Ontario Power Authority. During these events, citizens were encouraged to maximize their energy conservation activities. Smaller cities such as London, Guelph, Woodstock and Waterloo took part in the challenges annually. The final Voluntary Blackout Day was held on August 14, 2010.[31]

See also

- *Brittle Power*
- List of power outages
- New York City blackout of 1977

References

[1] "What Caused the Power Blackout To Spread So Widely and So Fast?" (http://genscape.com/pages.php?uid=4&sid=10&nid=34&act=news). Genscape. 2003-08-15. . Retrieved 2011-07-13.

[2] [articles.cnn.com/2003-08-14/us/power.outage_1_outage-power-plant-lightning-strike?_s=PM:US "Major power outage hits New York, other large cities"]. articles.cnn.com/2003-08-14/us/power.outage_1_outage-power-plant-lightning-strike?_s=PM:US. Retrieved 04/06/2012.

[3] "The great 2003 North America blackout" (http://archives.cbc.ca/science_technology/energy_production/clips/13545/). CBC. 2003-08-14. . Retrieved 2008-09-19.

[4] "Timeline" (http://www.cbc.ca/news/background/poweroutage/timeline.html). *INDEPTH: POWER OUTAGE* (CBC News). 2004-08-18. . Retrieved 2008-09-15.

[5] "Interim Report on the August 14, 2003, Blackout" (http://www.hks.harvard.edu/hepg/Papers/NYISO.blackout.report.8.Jan.04.pdf) (PDF). New York Independent System Operator. 2004-01-08. . Retrieved 2008-09-16.

[6] "Major power outage hits New York, other large cities" (http://www.cnn.com/2003/US/08/14/power.outage/). CNN. August 14, 2003. . Retrieved 2008-09-16.

[7] "The Costs Of Losing Our Night Skies" (http://www.illinoislighting.org/loss.html). www.illinoislighting.org. 2003-08-14. . Retrieved 2012-06-16.

[8] "Blackout 2003: Bush Rules Out Terrorism, Pledges Grid Review" (http://www.ens-newswire.com/ens/aug2003/2003-08-15-03.html). Newswire. 2003-08-15. . Retrieved 2011-07-20.

[9] "Canada gets blackout blame: World: News" (http://www.news24.com/News24/World/News/0,,2-10-1462_1402915,00.html). News24. 2003-08-15. . Retrieved 2010-04-21.

[10] Michehl R. Gent. "Statement of Michehl R. Gent President and CEO" (http://wayback.archive.org/web/20030815000000*/http://www.nerc.com/pub_doc/media-statement-08-16-03.doc). North American Electric Reliability Council. Archived from the original (http://www.nerc.com/pub_doc/media-statement-08-16-03.doc) on 18 August 2003. . Retrieved 31 March 2012.

[11] Mike Danforth and Ian Chillag (14 September 2012). "Episode 75: Strikes and Alibis" (http://howtodoeverything.org/post/31534266259/the-origin-of-strike-and-creating-a-cover-story). *How To Do Everything* (Podcast). NPR. . Retrieved 18 September 2012.

[12] "DOE chief sees no blackout penalty for FirstEnergy" (http://web.archive.org/web/20040224080845/http://www.forbes.com/markets/newswire/2003/11/19/rtr1153863.html). Forbes.com. 19 November 2003. Archived from the original (http://www.forbes.com/markets/newswire/2003/11/19/rtr1153863.html) on 24 February 2004. . Retrieved 31 March 2012.

[13] "U.S.-Canada Power System Outage Task Force Final Report" (https://reports.energy.gov/BlackoutFinal-Web.pdf) (PDF). . Retrieved 2012-06-16.

[14] Kevin Poulsen (2004-02-11). "Software Bug Contributed to Blackout" (http://www.securityfocus.com/news/8016). Securityfocus.com. . Retrieved 2010-04-21.

[15] Kevin Poulsen (2004-04-07). "Tracking the blackout bug" (http://www.securityfocus.com/news/8412). Securityfocus.com. . Retrieved 2010-04-21.

[16] "Microsoft Word - NERC Report on Blackout - Master.doc" (http://www.nerc.com/docs/docs/blackout/NERC_Final_Blackout_Report_07_13_04.pdf) (PDF). . Retrieved 2012-06-16.

[17] Lerner, Eric J. (October/November 2003). "What's wrong with the electric grid?" (http://www.tipmagazine.com/tip/INPHFA/vol-9/iss-5/p8.html). *The Industrial Physicist* (American Institute of Physics). . Retrieved 2009-06-10.

[18] "Chapter 5: How and Why the Blackout Began in Ohio" (http://www.nerc.com/docs/docs/blackout/ch5.pdf). NERC Final Report. .

[19] CNN: A timeline of the 2003 blackout (http://www.cnn.com/2003/US/08/16/blackout.chron.ap/index.html) (August 16, 2003)

[20] Coyne, John P. (2003-08-24). "Boom signaled power-line arc in Walton Hills" (http://web.archive.org/web/20071217151746/http://www.cleveland.com/blackout/index.ssf?/blackout/more/1061717873203660.html). *The Plain Dealer*. Archived from the original (http://www.cleveland.com/blackout/index.ssf?/blackout/more/1061717873203660.html) on 2007-12-17. . Retrieved 2011-08-30.

[21] "Looking back at the 2003 blackout" (http://eioc.pnl.gov/research/2003blackout.stm). Pacific Northwest National Laboratory. . Retrieved 2010-02-23.

[22] "CBC News Indepth: Power outage" (http://www.cbc.ca/news/background/poweroutage/numbers.html). Cbc.ca. . Retrieved 2010-04-21.

[23] The A&E reality television series *Airline* episode titled "Beyond Their Control" features Southwest employees dealing with the flight interruptions caused by the blackout.

[24] "Blackout Stops Presses, Shuts Web Sites," *NewsInc*, August 18, 2003

[25] Rick Lindquist, N1RL (October 2003). *Hams a Bright Spot during Power Blackout. QST*

[26] "Update on Fire and Explosion at Petro-Canada Lubricants Plant in Mississauga, Ontario" (http://web.archive.org/web/20071015182630/http://www.petro-canada.ca/en/media/1886.aspx?id=630421). Petro-Canada. 22 August 2003. Archived from the original (http://www.petro-canada.ca/en/media/1886.aspx?id=630421) on 15 October 2007. . Retrieved 31 March 2012.

[27] Carbon Disclosure Project (CDP) Greenhouse Gas Emissions Questionnaire (http://www.esmeefairbairn.org.uk/docs/cdp_report3.pdf)

[28] http://www.FDNewYork.com/lightsout.asp

[29] Greeno, Cherri (2003-08-25). "The Record.com" (http://pqasb.pqarchiver.com/therecord/access/647946491.html?dids=647946491:647946491&FMT=ABS&FMTS=ABS:FT). Pqasb.pqarchiver.com. . Retrieved 2010-04-21.

[30] "Casa Grande Valley Newspapers Inc" (http://www.zwire.com/site/news.cfm?newsid=10019242&BRD=1817&PAG=461&dept_id=68561&rfi=6). Zwire.com. 2010-03-20. . Retrieved 2010-04-21.

[31] "Blackout Day fades to black" (http://www.woodstocksentinelreview.com/ArticleDisplay.aspx?e=3231415). woodstocksentinelreview.com. 2011-10-27. .

External links

News stories

- CBC News (http://www.cbc.ca/news/background/poweroutage/) (in-depth feature)
- NPR News (http://www.npr.org/features/feature.php?wfId=1396052) (special feature)
- Ronda Hauben: Why Was the Blackout So Widespread? (http://www.heise.de/tp/english/inhalt/co/15464/1.html) (August 19, 2003)
- Dar Al Hayat: Al Qaeda claims responsibility (http://english.daralhayat.com/arab_news/08-2003/Article-20030818-14bdd659-c0a8-01ed-0079-6e1c903b7552/story.html) (August 18, 2003)
- San Francisco Chronicle: Chaos theories calculate the vulnerability of megasystems (http://sfgate.com/cgi-bin/article.cgi?f=/c/a/2003/08/15/MN191082.DTL) (August 15, 2003)
- BBC: Blackouts cause North America chaos (http://news.bbc.co.uk/1/hi/world/americas/3152451.stm) (August 15, 2003)
- CBC: Eastern blackout slowly lifting (http://www.cbc.ca/stories/2003/08/14/power_030814) (August 15, 2003)
- CNN: Major power outage hits New York, other large cities (http://www.cnn.com/2003/US/08/14/power.outage/index.html) (August 14, 2003)
- CBC: Blackout report blames Ohio utility (http://www.cbc.ca/stories/2003/11/19/blackout031119) (November 19, 2003)
- Towering Design Flaws (http://www.rmi.org/images/PDFs/EnergySecurity/E03-06_TowerDsnFlaws.pdf)
- USA TODAY: U.S. power grid in better shape 5 years after blackout (http://www.usatoday.com/money/industries/energy/2008-08-12-blackout-power-outage_N.htm) (August 12, 2008)

Other links

- Coverage and pictures (http://www.gothamist.com/archives/2003/08/15/the_new_york_city_blackout_edition.php)
- Map of outages (requires Flash) (http://www.usatoday.com/news/graphics/powerfailure/flash.htm)
- BBC: Share your experiences (http://news.bbc.co.uk/1/hi/talking_point/3152557.stm)
- Blackout History Project (http://blackout.gmu.edu/)
- Nine-Mile Point (http://www.eia.doe.gov/cneaf/nuclear/page/at_a_glance/reactors/nine_mile.html) and Fitzpatrick (http://www.eia.doe.gov/cneaf/nuclear/page/at_a_glance/reactors/fitzpatrick.html) reactors (EIA)
- US Canada Power System Outage Task Force Final Report (https://reports.energy.gov/)
- Blackout Analysis (http://www.io.com/~wazmo/blog/archives/2004_04.html)
- NYISO Final Report on the Blackout (http://www.nyiso.com/public/webdocs/newsroom/press_releases/2005/blackout_rpt_final.pdf) February 2005 (PDF)

- Blackout Day (http://www.blackoutday.ca/index.php) The Ontario energy conservation initiative in commemoration of the event.
- (http://ewh.ieee.org/soc/pes/psdpc/documents_blackout_panel/Blackout_White_Paper_By_IEEE_PES_AdComm_FINAL_082504.pdf)

North_American_Electric_Reliability_Corporation

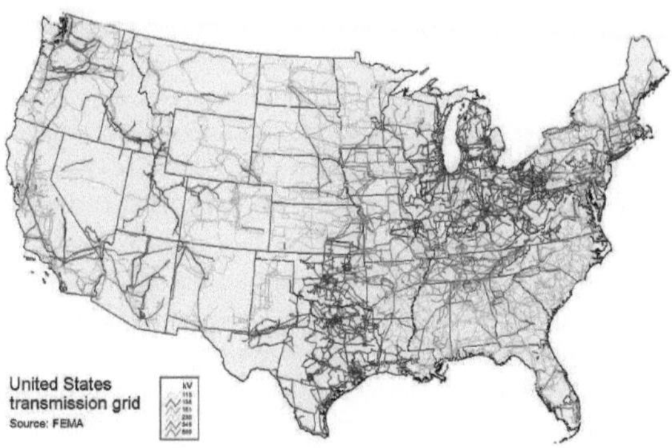

The contiguous United States power transmission grid consists of 300,000 km of lines operated by 500 companies

The **North American Electric Reliability Corporation (NERC)**, a nonprofit corporation based in Atlanta, Georgia, was formed on March 28, 2006, as the successor to the North American Electric Reliability Council (also known as NERC). The original NERC was formed on June 1, 1968, by the electric utility industry to promote the reliability and adequacy of bulk power transmission in the electric utility systems of North America. NERC's mission states that it is to "ensure the reliability of the North American bulk power system."[1]

NERC oversees eight regional reliability entities and encompasses all of the interconnected power systems of the contiguous United States, Canada and a portion of Baja California in Mexico.

NERC's major responsibilities include working with all stakeholders to develop standards for power system operation, monitoring and enforcing compliance with those standards, assessing resource adequacy, and providing educational and training resources as part of an accreditation program to ensure power system operators remain qualified and proficient. NERC also investigates and analyzes the causes of significant power system disturbances in order to help prevent future events.

NERC also provides for critical infrastructure protection (**NERC CIP**).

Origins of NERC

Early electric power systems, such as those installed by George Westinghouse and Thomas Edison, prior to the start of the 20th century were isolated central stations which served small pockets of customers independently of each other. As some of these power systems grew to cover larger geographic areas, it became possible to connect previously isolated systems. This allowed neighboring systems to share generation and voltage stability resources, providing mutual benefit to each side. However, tying power systems together with these early interconnections also introduced the risk that a single significant disturbance could collapse all of the systems tied to the interconnection. Generally it was decided that the benefits outweighed the risks, and by 1915 interconnections began to flourish and grow in size. By the end of the 1960s there were virtually no isolated power systems remaining in the lower forty-eight states and southern Canada; practically all power companies were attached to large interconnections.

In 1962, when the Eastern Interconnection was established in its current form, The Interconnected Systems Group (composed of Southern and Midwestern utility companies), the PJM Interconnection, and the Canada-United States

Eastern Interconnection (CANUSE) formed the Interconnection Coordination Committee to recommend an informal operations structure, which led to the formation of the North American Power Systems Interconnection Committee (NAPSIC). NAPSIC eventually grew to also include the Texas Interconnection and most of the companies in what is today the Western Electricity Coordinating Council (WECC), operating within the Western Interconnection.

On November 9, 1965, a relatively minor system disturbance triggered a power system protection component that was not properly configured. The interconnection was operating near peak capacity due to the extreme cold weather and high heating demand, and was therefore more vulnerable than usual. The small initial outage quickly cascaded into the Northeast Blackout of 1965. This disturbance revealed the extent that interconnections had evolved without adequate high-level planning and operating oversight to try to prevent such events, and that interconnected power systems frequently had varying operating standards and procedures developed somewhat independently by each member on the interconnection. Restoration efforts were also partially hampered due to the lack of common practices and coordination procedures. Furthermore, power system protection schemes were often designed with only a local power system's design in mind, meaning that they might misoperate in response to protection schemes activating in neighboring systems. This disturbance revealed the necessity to develop common operating and protection standards as well as plans to effectively coordinate power system restoration efforts.

The [Electric Reliability Act of 1967] proposed the creation of a council on power coordination. Although not enacted, the proposed legislation stimulated the development of an industry reliability council. The Federal Power Commission (predecessor of the Federal Energy Regulatory Commission) recommended the formation of a council on power coordination made up of representatives from each of the nation's regional coordinating organizations, to exchange and disseminate information and to review, discuss and assist in resolving interregional coordination matters.

The National Electric Reliability Council (NERC) was established by the electric utility industry, in response to the 1965 blackout. Nine regional reliability organizations were formalized under NERC. Also formalized were regional planning coordination guides, which NERC maintained. NAPSIC operations criteria and guides continued to be maintained and practiced voluntarily.

Although significant disturbances continued to occasionally occur, such as the New York City blackout of 1977, NERC undoubtedly played a significant role in minimizing the impact and frequency of these events. It is difficult to quantify this success because it is impossible to know how many disturbances were prevented by the influence of NERC and the reliability councils.

NERC today

Out of its long history, NERC developed a complex committee structure which brings together hundreds of industry expert volunteers in nearly 50 committees, sub-committees, task forces, and working groups considering issues from wind and renewable power integration to education to demand-side management and energy efficiency. NERC's role in raising awareness of reliability issues and creating the impetus to address them is intended to improve reliability every day.

With the passage of the Energy Policy Act of 2005, an "Electric Reliability Organization" was created to develop and enforce compliance with mandatory reliability standards in the U.S. This non-governmental, "self-regulatory organization" was created in recognition of the interconnected and international nature of the bulk power grid. In 2006, NERC applied for and was granted this designation.

Today, NERC's standards are mandatory and enforceable throughout the 50 United States and several provinces in Canada. Entities in the U.S. found to be in violation of a standard can be subject to fines of up to $1 million per day per violation.

Interconnections and Reliability Councils

Major interconnections

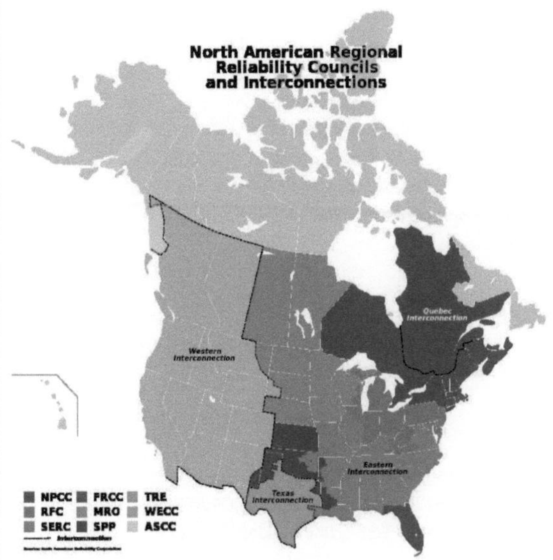

The two major and three minor NERC Interconnections, and the nine NERC Regional Reliability Councils.

- The Eastern Interconnection covers most of eastern North America, extending from the foot of the Rocky Mountains to the Atlantic seaboard, excluding most of Texas. The Eastern Interconnection is tied to the Western Interconnection via high voltage DC transmission facilities and also has ties to non-NERC systems in northern Canada. The reliability councils within the Eastern Interconnection are:
  - Florida Reliability Coordinating Council (FRCC)
  - Midwest Reliability Organization (MRO)
  - Northeast Power Coordinating Council (NPCC)
  - ReliabilityFirst Corporation (RFC)
  - SERC Reliability Corporation (SERC)
  - Southwest Power Pool, Inc. (SPP)
- The Western Interconnection covers most of western North America, from the Rocky Mountains to the Pacific coast. It is tied to the Eastern Interconnection at six points, and also has ties to non-NERC systems in northern Canada and Northwestern Mexico. The reliability council for the Western Interconnection is:
  - Western Electricity Coordinating Council (WECC)

Minor interconnections

- The Texas Interconnection covers most of the state of Texas. It is tied to the Eastern Interconnection at two points, and also has ties to non-NERC systems in Mexico. The reliability council for the Texas Interconnection is:
  - Electric Reliability Council of Texas (ERCOT)
- The Québec Interconnection covers the province of Québec and is tied to the Eastern Interconnection at two points. Despite being a functionally separate interconnection, the Québec Interconnection is often considered to be part of the Eastern Interconnection. The reliability council for the Québec Interconnection is:
  - Northeast Power Coordinating Council (NPCC)
- The Alaska Interconnection covers a portion of the state of Alaska and is not tied to any other interconnections. Due to its isolated nature, the Alaska Interconnection is not generally counted among North America's interconnections. The reliability council for the Alaska Interconnection is:
  - Alaska Systems Coordinating Council (ASCC), an affiliate member of NERC

NERC Authority

As part of the fallout of the Northeast Blackout of 2003, the Energy Policy Act of 2005 authorized the Federal Energy Regulatory Commission (FERC) to designate a national Electric Reliability Organization (ERO). On July 20, 2006, FERC issued an order certifying NERC as the ERO for the United States. Prior to being the National ERO, NERC's guidelines for power system operation and accreditation were referred to as *Policies*, for which compliance was strongly encouraged yet ultimately voluntary. NERC has worked with all stakeholders over the past several years to revise its Policies into *Standards*, and now has authority to enforce those standards on power system entities operating in the United States, as well as several provinces in Canada, by way of significant financial penalties for noncompliance. Efforts between NERC and the Canadian and Mexican governments are underway to obtain comparable authority for NERC to enforce its standards on the NERC member systems residing outside of the United States.

Cyber warfare

In April 2009, NERC issued a public notice that warns that the Electrical Grid is not adequately protected from cyber-warfare.[2]

See also

- Critical infrastructure protection
- Federal Energy Regulatory Commission (FERC)
- NERC Tag Energy Transaction tracking application
- V2G

References

[1] http://www.nerc.com/
[2] NERC Public Notice (http://online.wsj.com/public/resources/documents/CIP-002-Identification-Letter-040609.pdf)

External links

- Official website (http://www.nerc.com)
- List of NERC Regional Reliability Councils (http://www.nerc.com/page.php?cid=3|23)
- Reliability Report Warns of Transmission Needs with Wind Power Booming (http://apps1.eere.energy.gov/news/news_detail.cfm/news_id=12112) (USDOE)

Remote_Terminal_Unit

A **remote terminal unit (RTU)** is a microprocessor-controlled electronic device that interfaces objects in the physical world to a distributed control system or SCADA (supervisory control and data acquisition system) by transmitting telemetry data to the system, and by using messages from the supervisory system to control connected objects.[1] Another term that may be used for RTU is **remote telemetry unit**, the common usage term varies with the application area generally.

Architecture and communications

An RTU monitors the field digital and analog parameters and transmits data to the Central Monitoring Station. An RTU can be interfaced with the Central Station with different communication media (usually serial (RS232, RS485, RS422) or Ethernet). RTU can support standard protocols (Modbus, IEC 60870-5-101/103/104, DNP3, IEC 60870-6-ICCP, IEC 61850 etc.) to interface any third party software. In some control applications, RTUs drive high current capacity relays to a digital output (or "DO") board to switch power on and off to devices in the field. The DO board switches voltage to the coil in the relay, which closes the high current contacts, which completes the power circuit to the device. An RTU can monitor analog inputs of different types including 4 to 20 milliamperes (4–20 mA), 0–10 V., –2.5 to 2.5 V, 1–5 V etc.; the RTU or host system then translates this raw data into the appropriate units such as gallons of water left or temperature before presenting the data to the user via the HMI.

RTUs differ from programmable logic controllers (PLCs) in that RTUs are more suitable for wide geographical telemetry, often using wireless communications, while PLCs are more suitable for local area control (plants, production lines, etc.) where the system utilizes physical media for control. The IEC 61131 programming tool is more popular for use with PLCs, while RTUs often use proprietary programming tools.

Software and logic control

Modern RTUs are usually capable of executing simple programs autonomously without involving the host computers of the DCS or SCADA system to simplify deployment, and to provide redundancy for safety reasons. An RTU in a modern water management system will typically have code to modify its behavior when physical override switches on the RTU are toggled during maintenance by maintenance personnel. This is done for safety reasons; a miscommunication between the system operators and the maintenance personnel could cause system operators to mistakenly enable power to a water pump when it is being replaced, for example.

Comparison with other control systems

RTUs, PLCs and DCS are increasingly beginning to overlap in responsibilities, and many vendors sell RTUs with PLC-like features and vice versa. The industry has standardized on the IEC 61131-3 functional block language for creating programs to run on RTUs and PLCs, although nearly all vendors also offer proprietary alternatives and associated development environments.

In addition, some vendors now supply RTUs with comprehensive functionality pre-defined, sometimes with PLC extensions and/or interfaces for configuration.

Some suppliers of RTUs have created simple graphical user interfaces GUI to enable customers to configure their RTUs easily. In some applications dataloggers are used in similar applications.

A programmable automation controller (PAC) is a compact controller that combines the features and capabilities of a PC-based control system with that of a typical PLC. PACs are deployed in SCADA systems to provide RTU and PLC functions. In many electrical substation SCADA applications, "distributed RTUs" use information processors or station computers to communicate with digital protective relays, PACS, and other devices for I/O, and communicate with the SCADA master in lieu of a traditional RTU.

Applications

- Oil and gas remote instrumentation monitoring, (offshore platforms, onshore oil wells).
- Networks of remote pump stations (wastewater collection, or for water supply).
- Hydro-graphic monitoring and control (water supply, reservoirs, sewerage systems).
- Environmental monitoring systems (pollution, air quality, emissions monitoring).
- Mine site monitoring applications.
- Protection supervision and data logging of Power transmission network
- Air traffic equipment such as navigation aids (DVOR, DME, ILS and GP)
- Outdoor warning sirens, in both controlling them, and sending back data for verification of activation, trouble with the siren, or other data..

See also

- Telemetry

References

[1] Gordon R. Clarke, Deon Reynders, Edwin Wright, *Practical modern SCADA protocols: DNP3, 60870.5 and related systems* Newnes, 2004 ISBN 0-7506-5799-5 pages 19-21

Digital_protective_relay

Used in industrial process or control systems, a **digital protective relay** uses a microcontroller with software-based protection algorithms for the detection of electrical or process[1] faults. Such relays are also termed as microprocessor type protective relays.

Description and definition

The digital protective relay, or numeric relay, is a protective relay that uses a microprocessor to analyze power system voltages, currents or other process quantities for the purpose of detection of faults in an industrial process system.

The relay applies A/D (analog-to-digital) conversion processes to the incoming voltages, currents or other quantities to be monitored. The relay then analyzes the resultant A/D converter outputs to determine if action is required under its protection algorythm(s). As a minimum, magnitude of the incoming quantity, commonly using Fourier transform concepts (RMS and some form of averaging) would be used in a simple relay function. More advanced analysis can be used to determine phase angles, power, reactive power, impedance, waveform distortion, and other complex quantities. The relay is capable of applying advanced logic. It is capable of analyzing whether the relay should trip or restrain from tripping based on parameters set by the user, compared against many functions of its analogue inputs, relay contact inputs, timing and order of event sequences. The logic is user-configurable and can vary from simply changing front panel switches or moving of circuit board jumpers to accessing the relay's internal parameter setting webpage via communications link on another computer hundreds of kilometres away.

Digital relays often have some form of event recording. The event recording would include some means for the user to see the timing of key logic decisions, relay I/O (input/output) changes, and see, in an oscillographic fashion, at least the fundamental component of the incoming analogue parameters.

The relay may have an extensive collection of settings, beyond what can be entered via front panel knobs and dials, and these settings are transferred to the relay via an interface with a PC (personal computer), and this same PC interface may be used to collect event reports from the relay.

Digital/numerical relays also provides LCD Display, or display on a terminal through serial interface. This is used to display current/voltage values in real-time, and relay settings etc. More complex digital relays will have metering and communication protocol ports, allowing the relay to become an element in a SCADA system. Communication ports may include RS232/RS485 or Ethernet (copper or fibre-optic). Communication languages may include Modbus, DNP3 or IEC61850 protocols. By contrast, an electromechanical protective relay converts the voltages and currents to magnetic and electric forces and torques that press against spring tensions in the relay. The tension of the spring and taps on the electromagnetic coils in the relay are the main processes by which a user sets such a relay. In a solid state relay, the incoming voltage and current waveforms are monitored by analog circuits, not recorded or digitized. The analog values are compared to settings made by the user via potentiometers in the relay, and in some case, taps on transformers.

In some solid state relays, a simple microprocessor does some of the relay logic, but the logic is fixed and simple. For instance, in some time overcurrent solid state relays, the incoming AC current is first converted into a small signal AC value, then the AC is fed into a rectifier and filter that converts the AC to a DC value proportionate to the AC waveform. An op-amp and comparator is used to create a DC that rises when a trip point is reached. Then a relatively simple microprocessor does a slow speed A/D conversion of the DC signal, integrates the results to create the time-overcurrent curve response, and trips when the integration rises above a setpoint. Though this relay has a microprocessor, it lacks the attributes of a digital/numeric relay, and hence the term "microprocessor relay" is not a clear term.

History

The digital/numeric relay was introduced in the early 1980s, with AREVA, ABB Group's forerunners and SEL making some of the early market advances in the arena, but the arena has become crowded today with many manufacturers. In transmission line and generator protection, by the mid-1990s the digital relay had nearly replaced the solid state and electromechanical relay in new construction. In distribution applications, the replacement by the digital relay proceeded a bit more slowly. While the great majority of feeder relays in new applications today are digital, the solid state relay still sees some use where simplicity of the application allows for simpler relays, which allows one to avoid the complexity of digital relays.

Basic principles

Low voltage and low current signals (i.e., at the secondary of a voltage transformers and current transformers) are brought into a low pass filter that removes frequency content above about 1/3 of the sampling frequency (a relay A/D converter needs to sample faster than 2x per cycle of the highest frequency that it is to monitor). The AC signal is then sampled by the relay's analog to digital converter from 4 to 64 (varies by relay) samples per power system cycle. In some relays, a short history of the entire sampled data is kept for oscillographic records. Only the fundamental component is needed for most protection algorithms, unless a high speed algorithm is used that uses subcycle data to monitor for fast changing issues. The sampled data is then passed through a low pass filter that numerically removes the frequency content that is above the fundamental frequency of interest (i.e., nominal system frequency), and uses Fourier transform algorithms to extract the fundamental frequency magnitude and angle. Next the microprocessor passes the data into a set of protection algorithms, which are a set of logic equations in part designed by the protection engineer, and in part designed by the relay manufacturer, that monitor for abnormal conditions that indicate a fault. If a fault condition is detected, output contacts operate to trip the associated circuit breaker(s).

Protective element types

Protective elements refer to the overall logic surrounding the electrical condition that is being monitored. For instance, a differential element refers to the logic required to monitor two (or more) currents, find their difference, and trip if the difference is beyond certain parameters. The term element and function are quite interchangeable in many instances.

For simplicity on one-line diagrams, the protection function is usually identified by an ANSI device number. In the era of electromechanical and solid state relays, any one relay could implement only one or two protective functions, so a complete protection system may have many relays on its panel. In a digital/numeric relay, many functions are implemented by the microprocessor programming. Any one numeric relay may implement one or all of these functions.

A listing of device numbers is found at ANSI Device Numbers. A summary of some common device numbers seen in digital relays is:

- 11 - Multifunction Device
- 21 - Impedance
- 24 - Volts/Hz
- 25 - Synchronizing
- 27 - Under Voltage
- 32 - Directional Power Element
- 46 - Negative Sequence Current
- 40 - Loss of Excitation
- 47 - Negative Sequence Voltage
- 50 - Instantaneous Overcurrent (N for neutral, G for ground current)
- 51 - Inverse Time Overcurrent (N for neutral, G from ground current)
- 59 - Over Voltage
- 62 - Timer
- 64 - Ground Fault (64F = Field Ground, 64G = Generator Ground)
- 67 - Directional Over Current (typically controls a 50/51 element)
- 79 - Reclosing Relay
- 81 - Under/Over Frequency
- 86 - Lockout Relay / Trip Circuit Supervision
- 87 - Current Differential (87L=transmission line diff; 87T=transformer diff; 87G=generator diff)

References

[1] "Schweitzer Programmable Automation Controller" (https://www.selinc.com/SEL-2411/). Schweitzer Engineering Laboratories. . Retrieved 21 November 2012.

See also

- Electric power transmission
- Polyphase system
- Electricity distribution
- Overhead powerline
- Power outage
- Three-phase electric power

External links

- List of manufacturers of protective relays (http://www.p-r-eng.com/industrylinks)

Data_logger

A **data logger** (also **datalogger** or **data recorder**) is an electronic device that records data over time or in relation to location either with a built in instrument or sensor or via external instruments and sensors. Increasingly, but not entirely, they are based on a digital processor (or computer). They generally are small, battery powered, portable, and equipped with a microprocessor, internal memory for data storage, and sensors. Some data loggers interface with a personal computer and utilize software to activate the data logger and view and analyze the collected data, while others have a local interface device (keypad, LCD) and can be used as a stand-alone device.

Data logger Cube storing technical and sensor data

Data loggers vary between general purpose types for a range of measurement applications to very specific devices for measuring in one environment or application type only. It is common for general purpose types to be programmable; however, many remain as static machines with only a limited number or no changeable parameters. Electronic data loggers have replaced chart recorder in many applications.

One of the primary benefits of using data loggers is the ability to automatically collect data on a 24-hour basis. Upon activation, data loggers are typically deployed and left unattended to measure and record information for the duration of the monitoring period. This allows for a comprehensive, accurate picture of the environmental conditions being monitored, such as air temperature and relative humidity.

The cost of data loggers has been declining over the years as technology improves and costs are reduced. Simple single channel data loggers cost as little as $25. More complicated loggers may costs hundreds or thousands of dollars.

Data formats

Standardisation of protocols and data formats has been a problem but is now growing in the industry and XML is increasingly being adopted for data exchange. The development of the Semantic Web and the Internet of Things is likely to accelerate this present trend.

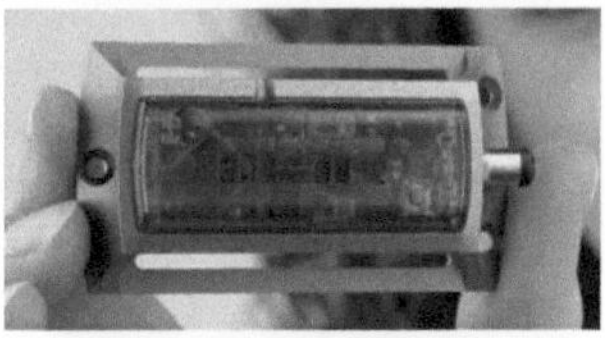

Small data logger with integrated sensors measuring temperature, pressure, humidity, light and 3-axis acceleration

Instrumentation protocols

Several protocols have been standardised including a smart protocol, SDI-12, that allows some instrumentation to be connected to a variety of data loggers. The use of this standard has not gained much acceptance outside the environmental industry. SDI-12 also supports multi drop instruments. Some datalogging companies are also now supporting the MODBUS standard. This has been used traditionally in the industrial control area, and there are many industrial instruments which support this communication standard. Another multi drop protocol which is now starting to become more widely used is based upon Canbus (ISO 11898). Some data loggers utilize a flexible scripting environment to adapt themselves to various non-standard protocols.

Data logging versus data acquisition

The terms data logging and data acquisition are often used interchangeably. However, in a historical context they are quite different. A data logger is a data acquisition system, but a data acquisition system is not necessarily a data logger.

- Data loggers typically have slower sample rates. A maximum sample rate of 1 Hz may be considered to be very fast for a data logger, yet very slow for a typical data acquisition system.
- Data loggers are implicitly stand-alone devices, while typical data acquisition system must remain tethered to a computer to acquire data. This stand-alone aspect of data loggers implies on-board memory that is used to store acquired data. Sometimes this memory is very large to accommodate many days, or even months, of unattended recording. This memory may be battery-backed static random access memory, flash memory or EEPROM. Earlier data loggers used magnetic tape, punched paper tape, or directly viewable records such as "strip chart recorders".
- Given the extended recording times of data loggers, they typically feature a mechanism to record the date and time in a timestamp to ensure that each recorded data value is associated with a date and time of acquisition in order to produce a sequence of events. As such, data loggers typically employ built-in real-time clocks whose published drift can be an important consideration when choosing between data loggers.
- Data loggers range from simple single-channel input to complex multi-channel instruments. Typically, the simpler the device the less programming flexibility. Some more sophisticated instruments allow for cross-channel computations and alarms based on predetermined conditions. The newest of data loggers can serve web pages, allowing numerous people to monitor a system remotely.
- The unattended and remote nature of many data logger applications implies the need in some applications to operate from a DC power source, such as a battery. Solar power may be used to supplement these power sources. These constraints have generally led to ensure that the devices they market are extremely power efficient relative to computers. In many cases they are required to operate in harsh environmental conditions where computers will not function reliably.
- This unattended nature also dictates that data loggers must be extremely reliable. Since they may operate for long periods nonstop with little or no human supervision, and may be installed in harsh or remote locations, it is imperative that so long as they have power, they will not fail to log data for any reason. Manufacturers go to great length to ensure that the devices can be depended on in these applications. As such dataloggers are almost completely immune to the problems that might affect a general-purpose computer in the same application, such as program crashes and the instability of some operating systems.

Applications

Applications of data logging include:

Data logger application for weather station at P2I LIPI

- Unattended weather station recording (such as wind speed / direction, temperature, relative humidity, solar radiation).
- Unattended hydrographic recording (such as water level, water depth, water flow, water pH, water conductivity).
- Unattended soil moisture level recording.
- Unattended gas pressure recording.
- Offshore buoys for recording a variety of environmental conditions.
- Road traffic counting.
- Measure temperatures (humidity, etc.) of perishables during shipments: Cold chain.[1]
- Process monitoring for maintenance and troubleshooting applications.
- Process monitoring to verify warranty conditions
- Wildlife research with pop-up archival tags
- Measure vibration and handling shock (drop height) environment of distribution packaging.[2]
- Tank level monitoring.
- Deformation monitoring of any object with geodetic or geotechnical sensors controlled by an automatic deformation monitoring system.
- Environmental monitoring.
- Vehicle Testing (including crash testing)
- Motor Racing
- Monitoring of relay status in railway signalling.
- For science education enabling 'measurement', 'scientific investigation' and an appreciation of 'change'
- Record trend data at regular intervals in veterinary vital signs monitoring.
- Load profile recording for energy consumption management.
- Temperature, Humidity and Power use for Heating and Air conditioning efficiency studies.
- Water level monitoring for groundwater studies.
- Digital electronic bus sniffer for debug and validation

Future directions

Data Loggers are changing more rapidly now than ever before. The original model of a stand alone data logger is changing to one of a device that collects data but also has access to wireless communications for alarming of events, automatic reporting of data and remote control. Data loggers are beginning to serve web pages for current readings, e-mail their alarms and FTP their daily results into databases or direct to the users.

Examples

- A flight data recorder (FDR), a piece of recording equipment used to collect specific aircraft performance data. The term may also be used, albeit less accurately, to describe the cockpit voice recorder (CVR), another type of data recording device found onboard aircraft.
- An event data recorder (EDR), a device installed by the manufacturer in some automobiles which collects and stores various data during the timeframe immediately before and after a crash.
- A voyage data recorder (VDR), a data recording system designed to collect data from various sensors on board a ship.

- Ultra Wideband Data Recorder, high-speed data recording up to 2 GigaSamples per second.
- The growing, preparation, storage and transportation of food. Data logger is generally used for data storage and these are small in size.
- In automobiles, all diagnostic trouble codes (DTCs) are logged in engine control units (ECUs) so that at the time of service of a vehicle, a service engineer will read all the DTCs using Tech-II or similar tools and will come to know problems occurred in the vehicle.
- In the racing industry, Data Loggers are used to record data such as braking points, lap/sector timing, and track maps etc. A product produced by Race Technology capable of doing this is the DL1 Data logger and can be used on any type of vehicle. There are also other alternative products available on the market.
- In embedded system and digital electronics design, specialized high-speed digital data logger help overcome the limitations of more traditional instruments such as the oscilloscope and the logic analyzer. The main advantage of a data logger is its ability to record very long traces, which proves very useful when trying to correct functional bugs that happen once in while.

See also

- Bus analyzer
- Computer data logging: logging APIs, server logs & syslog, web logging & web counters
- Continuous emissions monitoring system
- Runtime intelligence
- Sequence of events recorder
- SensorML
- Shock and vibration data logger
- Temperature data logger

References

[1] Riva, Marco; Piergiovanni, Schiraldi, Luciano; Schiraldi, Alberto (January 2001). "Performances of time-temperature indicators in the study of temperature exposure of packaged fresh foods". *Packaging Technology and Science* **14** (1): 1–39. doi:10.1002/pts.521.

[2] Singh, J; Singh, Burgess (2007). "Measurement, Analysis, and Comparison of the Parcel Shipping Shock and Drop Environment of the United States Postal Service with Commercial Carriers". *JOTE* **35** (3). doi:10.1520/JTE100787.

Article Sources and Contributors

Sequence_of_events_recorder *Source*: http://en.wikipedia.org/w/index.php?title=Sequence_of_events_recorder *Contributors*: GWS EE, Hooperbloob, Koavf, Malcolma, Obankston, RHaworth, SkeletorUK, Toloquta, Tony1, 3 anonymous edits

Slashme, Slightsmile, Spikebailey666, Starazagora, Steambt21, Tabletopper323, Tbonnema, Tesla1231, The Anome, Thequality, Tnxman307, Utuado, Velella, WhatamIdoing, Widr, William Flower, Ygfperson, 210 anonymous edits

JudeFawley, Kevin Saff, Krishnaeee, Kuru, Lobrieniii, M@sk, Martin Hollender, Mausy5043, Mav, Mbeychok, Mikewings, Miym, Mkill, Mydogategodshat, Niceguyedc, Overjive, Owenpr, Pjadot, Pjetter, Pnm, PotentialDanger, Psyllicon, Pugges, Qfli78, Radiojon, Rainor, Rajendu.choubisa, Reclute, RedWolf, Rembecki, Ripogenus77, Rjwilmsi, Robpelu, RoyKok, Royalguard11, Rprogatto, Russ84, Sadads, SamMason, Shrew, SirLamer, Smack, Spalding, Srikanthpisipati, Sun Creator, Supersteve04038, Textory, The Rambling Man, Tony1, TriniTriggs, User48365, Utsav91292, Vikasy, Wdsanderson, Welsh, WillemHazenberg, Wtshymanski, 211 anonymous edits

SCADA *Source*: http://en.wikipedia.org/w/index.php?title=SCADA *Contributors*: 1exec1, 84user, A3r0, AWoodland, Aboalbiss, Adolphus79, Afredman, Ahj, Ahmedlinux, Ahoerstemeier, Aidan64, Alan Liefting, AlexDovgan, Alexocon, Alperkaradas, Antiopi, Ap, Apapadop, Aphstein, Arthena, Arthur Rubin, Aviator67, BMaxfield, Back ache, Baka610, Balasana, Baltazuvedra, Bdc101, Beetstra, Bertus, Billymac00, Bjagpal, Bluemagoo, Bruyninc, Bsandg, Buhrerf, Camelek, CanisRufus, Chester Markel, Chill doubt, Chrisgraty13247, Cj, CoJaBo, Colin Pye, Collieman, Control.optimization, Crunchy Numbers, Cybercobra, D6, DBigXray, DRTllbrg, Dan Medbury, Danski14, Darp-a-parp, David Gerard, DavidCBryant, DeadEyeArrow, Deasington, Decisal, Diceman, Dougher, Dpc1962, Dreadstar, Dtroncho, Edward, EncMstr, Eskalin, Ewulp, Faganp, Femto, FeralOink, Fingers-of-Pyrex, Frank Lofaro Jr., Fromnewark, Fuzzybsc, GWS EE, Gary84, Gepwiki, Ghettoblaster, Ghostclay, Giftlite, Glennb28, Goatasaur, GoldenTorc, Gpvos, Gwinna, Gwpdispatcher, H.toraby, HaPi, Hb277, Hdmaclellan, Hegh, Heirpixel, Helios369, Herbythyme, Hu12, Inductiveautomation, Industrialautonet, Infoceantom, Infotek411, Inwind, Iridescent, Ishi Gustaedr, JForget, Jack Greenmaven, JackCoke, JanRaas, Jarsyl, Jean.julius, Jesper Krog, Jharistin, Jmgonzalez, Joantorres, John Frank Keczmerski, John Reed Riley, Jolastar, JonHarder, JorgeGG, Jsmaye, Jstplace, Juanparve, Kajano, KarlJorgensen, Kh2155, Kiltbear, Kinema, Kingpin13, Kjellgro, Kjkolb, KnightRider, Knowcow, Kotasik, Kozuch, Krashlandon, Krushia, Kuklojob, Kuru, Kylan, Ladybetty, Lambertk, Landon1980, Lehrig, Light current, Limtohhan, Lokeshchaudhari, M@sk, MCorley, MER-C, Magioladitis, Malafaya, Malo, Mandarax, Manik762007, Masgatotkaca, Mat i, Maurice Carbonaro, Mchartrand, Michael Devore, Microsp, Mikezuff, Moreto, Morning277, Mortense, MrOllie, Mtutin, Mwatwe01, Mydogategodshat, N5iln, Neddy53, Newkopit, Neworion, Neznanec, Ngpd, Nick Number, Ninly, Ninokurtalj, Nmnogueira, Oakb, Olarssen, Olivier, Ost316, Oyoyoy, Pavel Vozenilek, Peaceray, Pengo, PinchasC, Pjetter, Pnm, Pointe LaRoche, Public Menace, Pugges, Quuxplusone, Qxz, RedWolf, Requestion, Rick2378, Romal, Ronark, RoyKok, Rufanov, Rwwww, S3000, Sam Hocevar, Sat84, Sbacle, Scada-sys, Scada01, Scadateer, Schastain, SchreiberBike, Scott white85, Seaphoto, Sfax.tn, Shanes, SiobhanHansa, Skortier, Sole Soul, Squash1978, Stevecarsonr, Stevewrd, StickyWidget, Superdude99, Suruena, Takorman, Tassedethe, Techtonik, Textory, Tgeairn, The Anome, The Thing That Should Not Be, Thebaroque, Thrushwood, Thunderbird62, Tie.mudge, Tmaufer, Toh, Tomia, Tomkalayil, Tony1, Truthflux, Tsange, Tschreider, TwoTwoHello, Umeshktangnu, Uncle Dick, Unfree, UnitedStatesian, User A1, Utilitysupplies, Vir Novus, Vocaro, Voidxor, Vonkje, Vsmith, WCFrancis, Welsh, Wik, Wikieditoroftoday, Wknight94, Wmmorrow, Wolfgang S, Wtshymanski, XandroZ, XtraChewy0, Yaronf, Yesid8, Yonkie, Zek725, 662 anonymous edits

Programmable_logic_controller *Source*: http://en.wikipedia.org/w/index.php?title=Programmable_logic_controller *Contributors*: (aeropagitica), A876, Aaron Brenneman, Afshan5zeba, Ahoerstemeier, Alansohn, Ale jrb, Alexandermorgantop, Allecher, Alperkaradas, Amillar, Anaveenraj31, Anbu121, Andreas Kaufmann, Andy Dingley, AnthonyQBachler, Aotewell, ArséniureDeGallium, Athabaska-Clearwater, Attilios, Baa, Back ache, Baka610, Banaticus, Benandorsqueaks, Bin95, Bircwiki, BlueDevil, Bongwarrior, Bovineone, Bozoid, Brent s plc, Brian Radwell, Burpen, CIreland, Calmer Waters, Can't sleep, clown will eat me, CanOfWorms, CanadianLinuxUser, CanisRufus, Canoe1967, Cartiman, CesarB, Chamal N, Charles Sturm, ChemGardener, Chris.rickey, Christian List, Ciphers, Cjmulvey, Closedmouth, Cmichael, Collieman, Control.optimization, Conversion script, Credema, Crunchy Numbers, Cst17, Ctdmrod06010, D4g0thur, DARTH SIDIOUS 2, DJ LoPaTa, DabMachine, Dailynetworks, Darkwind, DeadEyeArrow, Deb, Debresser, Denisarona, DerHexer, Derectus, Diagonalfish, DieSwartzPunkt, Discospinster, Doug Coldwell, Dougher, Dougofborg, Dtrebbien, El C, Elmschrat, Email4mobile, Epbr123, Euryalus, Excirial, Fbolanos, Fieldday-sunday, Finalreminder, Frapacino, Fuzlyssa, Fæ, GRAHAMUK, Gail, Gary King, Gary2001, Gepwiki, Gerfriedc, Gg automation, Giftlite, Glenn, Golgofrinchian, GrimmReaperSound, Gstortz, Gurt Posh, HaPi, Hammertime, Harriv, Hooperbloob, Hughjack, Icestorm815, Ilyushka88, Injeek6, Ipso2, Iridescent, Ishi Gustaedr, Isis4563, IvanLanin, J.delanoy, JMiall, Jackfork, Jcipc2004, Jerry, JethroElfman, Jfdwolff, Jfmantis, Jim.henderson, Jim1138, JimScott, Joe.dolivo, John Vandenberg, John of Reading, Jorunn, Jpbowen, Jprg1966, JustAGal, Jwy, K4rmix, KD5TVI, KFullerton, Kakesson, Kbdank71, Kimchi.sg, Kku, Kntrabssi, Kralizec!, Krushia, Kuru, Ladybetty, Lakee911, Lambtron, Lcwk86, Ldwolk, LessHeard vanU, Lightmouse, Limebite, Lindsey Sh, Ljudina, Lmatt, Lonelydarksky, LorenzoB, Loupeter, Loyt, MCorley, MFH, ML5, Mabdul, Maddensue, MahdiEynian, Mahmdmuhy, Mandarax, Marek69, Mark91, MartyRobar, Martynas Patasius, Maryeputnam, Materialscientist, Maurice Carbonaro, Mdd, Meno25, Michael Hardy, Microsp, Midhart90, Mike Rosoft, Minna Sora no Shita, Mitchan, Mixabest, Mje112, MkClark, Moaz786, Modster, Monkan, Morphaeous, Mortense, MrBlueSky, MrOllie, MrRK, Mrand, Mydogategodshat, N5iln, Namoson, Neilyboy4306, NellieBly, Netbymatt, Newt Winkler, Nharipra, North8000, Novaseminary, Nubiatech, Nwk, OlEnglish, Oliver202, Ot, Oxymoron83, PLClogistics, Passion of Knowledge, Patden, Pax85, Pdcook, Pdfsupply, Pelcer, Pengo, Peter Flass, Petrb, PhilKnight, Philip Trueman, PierreAbbat, Plc training, Pointbonita, Poliyal, Ponyo, Pratik rathore, Prdejong, PurpleChez, RB972, Radwell International, Rahul.mishra, Raldrich123, Ray Van De Walker, Razorflame, RedWolf, Redvers, Requestion, Rich Farmbrough, Ripogenus77, Rjwilmsi, Rkinner, Robertvan1, Robofish, Rocketrod1960, Rolfds, Ronz, RoyKok, Ryan Vesey, SJP, SUL, Sanjivee, Sarg, Scantime, Sean D Martin, Seaphoto, Sevicke, Shai-kun, Shanes, Shariat ipe, Shible.isteak, Shirik, Shoeofdeath, Sidelight12, SiobhanHansa, Skarebo, Skipperalfie, Skyscrap27, Smhaatif, Smurrayinchester, Solarra, Some jerk on the Internet, Sonett72, Spalding, Srleffler, Ssd, Stan Shebs, SteffiKl, Stickee, Straight-shooter(42), Superluser, Supersteve04038, Suruena, TERdON, Tasnai, Tatau1234, TcomptonMA, Tech Jockey, Tech77, Teles, Texture, The Utahraptor, Theo10011, Think outside the box, Tho2468, Tholly, Tiptoety, Tobias Bergemann, Toffanin, Tom harrison, TomCerul, Tripledot, Turk oğlan, TutterMouse, Ugen64, Uncle Milty, User A1, Utcursch, V3co, VGarner, VMS Mosaic, Velella, Versus22, Vev, Vijayamurugan.p@gmail.com, Wdfarmer, Wernher, Wikipelli, Wildthing61476, Wimt, Woohookitty, Wtshymanski, XAVeRY, Xhantar, Yaamboo, YakiraLight, Yaronf, Yaseen805, Yosyk07, YouAndMeBabyAintNothingButCamels, Zalgo, ВикиКорректор, 1097 anonymous edits

Northeast_blackout_of_2003 *Source*: http://en.wikipedia.org/w/index.php?title=Northeast_blackout_of_2003 *Contributors*: 08OceanBeach SD, A2Kafir, ABCD, AHands, Adam Bishop, Adam12901, Adam850, Afeldman43, Aff123a, Afiler, Ahoerstemeier, Ahpook, Aivazovsky, Alaney2k, Alansohn, Albert109, Alex Ruddick, Alex.tan, Alex756, Altonbr, Anc516, Andre Engels, Andres, Andy G, Angr, AnonMoos, Ansabakhan, Antandrus, Anthony Appleyard, ArnoLagrange, Arx Fortis, Aryamehrr, Asmendel, AstroHurricane001, Astros4477, Audemus Defendere, BD2412, Babbler, Barberio, BarkerJr, BassPlyr23, Bdesham, Bearcat, Bellebutrfly03, Benjamintchip, Benkin, Betacommand, BillC, Birdhombre, Biruitorul, Bkell, Bkonrad, Blatantlysubtle, Bletch, Blueiris, Bobblehead, Bobblewik, Books305060, Branden, Brandon5485, Brendan2003, Brion VIBBER, Brun8, C777, CComMack, CSWarren, CStyle, Calliope nh, Camerafiend, Canderra, Captain Cornflake, CaribDigita, Carnildo, CastAStone, Cema, CesarB, Cg-realms, Chrisahn, Cimon Avaro, Circeus, Cjmarsicano, Cjmnyc, Clngre, CokeBear, Colipon, Colonies Chris, ColtsScore, CommonsDelinker, Confiteordeo, Creflo, Crosbiesmith, CrunchySkies, Cstaffa, CutOffTies, Cyan, Cyberia23, DOSGuy, DXRAW, Dale Arnett, Daniel C. Boyer, Daniel McMullen, Daniel Quinlan, Danno2530, Danorton, Darkman882005, Darrelljon, Darwindecks, David elliott lewis, DavidSSabb, Dawnseeker2000, Deathbal101, Deities, Denelson83, Dglynch, Digitalme, Dirtgirl, Dunezone, Dwheeler, Dynamite XI, E Pluribus Anthony redux, Ed Cormany, Ed Poor, Ed g2s, Edward, Eelamstylez77, Efghij, Eloquence, Emmbee, Enchanter, EoGuy, Epiovesan, Eratticus, Eric119, Eurosong, Evercat, Everything counts, Evil saltine, FD347, FironDraak, Firsfron, Fjbfour, Floydian, FlyHigh, Foiltape, Foobar, Fourthords, Franny Wentzel, Freedom24, Freerick, Fubar Obfusco, Fuhghettaboutit, Funandtrvl, GPHemsley, Gaius Cornelius, Galmicmi, GavinSharp, Geary, GeorgeLouis, Gerryschmitt, Gleemonex, Groyolo, Gujuguy, Gurch, Hbdragon88, HeadSnap, Herr Ratselhaft, Hugo999, Hurc, Hydrogen Iodide, Iansanderson, Impaciente, Imran, Indefatigable, IrishPete, Isopropyl, Ixfd64, J Clear, J JMesserly, Jadedcrypto, Jahiegel, Jamcib, Jamesday, Jasonp, Jaysbro, Jeff3000, Jengod, Jenks24, Jennifer Zeher, JeremyMcCracken, Jiang, Jim E. Black, Jim Tour, Jim.henderson, JimVC3, JimWae, Jimjohnnyc, Jmorgan, Jodan, John Quincy Adding Machine, Johnfos, Johnrpenner, Jok2000, Jonyyeh, Jstreutker, Jtkiefer, Juliancolton, JustAnMD, KD-G722, KJRehberg, KJS77, Kanabekobaton, Kchishol1970, Keeves, Kellee94, Kerowren, Kharker, King of Hearts, Klparrot, Kmeister, Koavf, Krupo, Kushal kumaran, Kw0134, LOL, Lapaz, Laug, Ldrhcp, LeedsKing, Lem33, Leventio, Lexor, Lightmouse, Ligtymn, Lir, Llb9977, Lokal Profil, Lowellian, M.S.K., MSTCrow, Macosta, Magnus Manske, Mahjong705, Mais oui!, Mapsax, Marcinjeske, Mardus, Marigold Sun, Mark K. Jensen, Markbarner, Marketdiamond, Markustwofour, MarnetteD, MasterNetHead, Matches10, MathiasRav, Matthew Woodcraft, Maury Markowitz, Mav, Maxim, Mblumber, Mdchachi, Menchi, Michael Devore, Michael Hardy, Miesianiacal, Miguel, Mindspillage, Minesweeper, Mintguy, Mishmash8, Misterkillboy, Mjs1991, Mkweise, Moncrief, Montrealais, MoraSique, Mr. Laser Beam, Naddy, Nanobug, NeonMerlin, Netalarm, Netesq, Newkai, Nick Number, Nickj, Nights Not End, Nightscream, Nneonneo, Noommos, NorthernThunder, Nostrum, Nyarpy, Oblivious, OhanaUnited, Omtay38, Orgreenism, Ornil, POVpushee, Pacific Coast Highway, Pakaran, Parsiferon, Pascal, Pascal666, Pats1, Paul Benjamin Austin, Paulmco, Paulwehr, Pb30, Phil Boswell, PhilthyBear, Pickles8, PickupmanUP1966, Pietrow, Pjb dinky, Pjf, Plasma east, Poor Yorick, Porqin, Professor water, Public Menace, Pwnage8, Quarl, Quicksilvre, Ragemanchoo, Ram4eva, Reaper Eternal, Red phase, RedWolf, Reddi, Reisio, Reynwah, Rhombus, Richard416282, RickK, Rico402, RingtailedFox, Rjairam, Rjwilmsi, Rmhermen, Rob T Firefly, Rodrigue, RoyBoy, SBArmstrong, SD6-Agent, SNIyer1, SNIyer12, SYSS Mouse, Sam Hocevar, Samuel, Saxbryn, SchuminWeb, Scobe3, Seattle Skier, Sgt Pinback, Shiftchange, Shultz, Sietse Snel, Sim, SimonP, Sk5893, Skabat169, Skootles, Skylap, SmthManly, Smurrayinchester, Snickerdo, Someone else, SpeakerFTD, Sprocket, Ssbohio, Steam5, Stuart P. Bentley, Substatique, Sully, TACD, THD3, TPIRFanSteve, TWCarlson, Tariqabjotu, Tassedethe, Telso, Teukros, The Anome, ThePedanticPrick, Theanphibian, Theresa knott, Thetorpedodog, Thinkpad, Threedots dead, Thumperward, Tiles, Tillwe, Tim!, Tinton5, Tinus, Tlonca, Tlotoxl, Tom, Tony Fox, Trontonian, Troy112233, Truthanado, Tweetylns, TwinsMetsFan, Twinsday, UtherSRG, Uvaduck, VBerrini, Valermos, Vancouverguy, Vargob, Varlaam, Vegaswikian, VeryVerily, Viajero, Vina, Vingt cent, Vzbs34, Wa2ise, WayKurat, Wayne Miller, Wayward, Webfan29, Wenteng, Wfeidt, Whkoh, WideArc, Wik, Wizardman, Wmahan, Wtshymanski, X4nd1, XxTrillvillexX9, Xxhopingtearsxx, Ylee, Yoho2001, YoungScholaratHeart553, Zanimum, ZeWrestler, Zephalis, Zombiejesus, Zzyzx11, 711 anonymous edits

North_American_Electric_Reliability_Corporation *Source*: http://en.wikipedia.org/w/index.php?title=North_American_Electric_Reliability_Corporation *Contributors*: Amartineau, Anurag05, Beagel, Beyond My Ken, Bouchecl, Catamorphism, CopperSquare, Deanlaw, Dgrinkev, DonBenjamin, Ed Poor, Fjbfour, Green Squares, Hforbes, Hmains, Hugo999, J JMesserly, Jason.straathof, Jbertsch, Linas, Locke411, Michael Hardy, Mrdice, Mrtrey99, Neilc, Poor Yorick, Ppelleti, Pradameinhoff, Rich Farmbrough, Rolonation, Tiles, Tperlin, Voidxor, Wdwd, WikHead, Yaki-gaijin, 27 anonymous edits

Remote_Terminal_Unit *Source*: http://en.wikipedia.org/w/index.php?title=Remote_Terminal_Unit *Contributors*: 171046, Afredman, Ajeetkrd, Anna Lincoln, Anomalocaris, AnonUser, Ashuwiki19, Azazyel, Bailz1, Beauwoods, Biblbroks, Bongwarrior, Bsandg, Bumm13, ChrisGualtieri, CoJaBo, Danger, DeXXus, Diamondland, Diceman, Discospinster, Dpolley, Dschmidt, Ece.kassem, Faught, GWS EE, Gene Nygaard, Guitouf38, H.toraby, Hu12, J04n, JasonCoutinho, Ji.rodriguezmarin, Johnhunt, JonHarder, Kajano, Kh2155, ML5, Materialscientist, Matusz, Michael Hardy, Mild Bill Hiccup, Ng.j, Padma1012, Peter Horn, Sonett72, Steinsky, Stevecarsonr, SuperMono, Swen, Tabletop, Thunderbird62, Tom beebe, Tomilaurell, Tony1, Tothwolf, Truthanado, Txomin, Valayogendra, Velella, Voidxor, WCFrancis, Wtshymanski, 112 anonymous edits

Digital_protective_relay *Source*: http://en.wikipedia.org/w/index.php?title=Digital_protective_relay *Contributors*: Ansabakhan, ChrisGualtieri, CommonsDelinker, Crystallina, Dougher, Frmorrison, GWS EE, Hooperbloob, Igtorque, Jgpinheiro, JohnHorak, Kylan, LittleWink, Mathonius, Oli Filth, Pietrow, Pmetzger, Running Will, Rwand, S H Richards, Shrike, Throwaway85, Toffile, Vegaswikian, Will Beback Auto, Wtshymanski, Yworo, 54 anonymous edits

Data_logger *Source*: http://en.wikipedia.org/w/index.php?title=Data_logger *Contributors*: 171046, 240techie, Anakwisnu, Barry abs, Blainster, BreezeSEO, BrianMcM, Brucedenney, C00ch, Campbellsci.com, Canuckian89, Cburnett, ChrisGualtieri, Cntras, Colonies Chris, Ctbolt, Curps, Danno uk, Darkwind, Ddevil226, Deb, Debresser, Deli nk, Deville, Dirkjot, Elmschrat, Elubofsky, Engineer1122, Epbr123, Eric Hegi, Erik1980, Fdixon, Finejon, Frap, Hankwang, Heron, Hooperbloob, IgorBGD, Inc ru, Interiot, Jcrm, Johnrcrellin, Jomodano, Jthomas999, Julia W, Karnesky, Kuru, Ladybetty, Lear's Fool, Levineps, MER-C, MSRDatenlogger, Materialscientist, Mbc362, Michael.Sage, Modiste, MrOllie, Mrthumperwabbit, Mwanner, Mwtoews, Nickel33, Nono64, Obankston, Pax:Vobiscum, Pearle, Philip Trueman, Pinethicket, Pkgx, Pochi1996, Prunesqualer, Q509, Quantumobserver, Radagast83, Reinyday, Rlsheehan, Roberteleejr, Rogerfrost, Rvoorhees, SJP, Sasan.j, Soyourowned, SuperMono, Tasc, TempLogInforPageMove, The flamingo, Tresiden, TwistedGray, Tyson882, Velella, Vsmith, Wiki Wiki Will C, Wtshymanski, Zoicon5, 179 anonymous edits

Image Sources, Licenses and Contributors

File:GPS Satellite NASA art-iif.jpg *Source*: http://en.wikipedia.org/w/index.php?title=File:GPS_Satellite_NASA_art-iif.jpg *License*: unknown *Contributors*: Bricktop, Bryan Derksen, Common Good, Edward, GDK, Jafeluv, Michaelfavor, Túrelio, 13 anonymous edits

File:Magellan GPS Blazer12.jpg *Source*: http://en.wikipedia.org/w/index.php?title=File:Magellan_GPS_Blazer12.jpg *License*: unknown *Contributors*: User:Nachoman-au

File:KyotoTaxiRide.jpg *Source*: http://en.wikipedia.org/w/index.php?title=File:KyotoTaxiRide.jpg *License*: unknown *Contributors*: Paul Vlaar

File:GPS on smartphone cycling.JPG *Source*: http://en.wikipedia.org/w/index.php?title=File:GPS_on_smartphone_cycling.JPG *License*: unknown *Contributors*: User:HawaiianMama

File:2 SOPS space systems operator 040205-F-0000C-001.jpg *Source*: http://en.wikipedia.org/w/index.php?title=File:2_SOPS_space_systems_operator_040205-F-0000C-001.jpg *License*: unknown *Contributors*: United States Air Force photo by Airman 1st Class Mike Meares)

Image:NAVSTAR GPS logo shield-official.jpg *Source*: http://en.wikipedia.org/w/index.php?title=File:NAVSTAR_GPS_logo_shield-official.jpg *License*: unknown *Contributors*: Koavf, Michaelfavor, Minna Sora no Shita, MithrandirMage, Tsaitgaist, 5 anonymous edits

Image:50th Space Wing.png *Source*: http://en.wikipedia.org/w/index.php?title=File:50th_Space_Wing.png *License*: unknown *Contributors*: United States Air Force

File:Global Positioning System satellite.jpg *Source*: http://en.wikipedia.org/w/index.php?title=File:Global_Positioning_System_satellite.jpg *License*: unknown *Contributors*: User:SCEhardt

File:GPS monitor station.jpg *Source*: http://en.wikipedia.org/w/index.php?title=File:GPS_monitor_station.jpg *License*: unknown *Contributors*: Bubba73 (You talkin' to me?), (Jud McCranie). Original uploader was Bubba73 at en.wikipedia

File:GPS Receivers.jpg *Source*: http://en.wikipedia.org/w/index.php?title=File:GPS_Receivers.jpg *License*: unknown *Contributors*: Bdk, Head, Jensens, Lzur, MB-one, Partyzan XXI, Ustas, WikipediaMaster, 1 anonymous edits

File:j32 1 small.jpg *Source*: http://en.wikipedia.org/w/index.php?title=File:J32_1_small.jpg *License*: unknown *Contributors*: User:Beao

File:GPS roof antenna dsc06160.jpg *Source*: http://en.wikipedia.org/w/index.php?title=File:GPS_roof_antenna_dsc06160.jpg *License*: unknown *Contributors*: User:David.Monniaux

File:US Navy 030319-N-4142G-020 Ordnance handlers assemble Joint Direct Attack Munition (JDAM) bombs in the forward mess decks.jpg *Source*: http://en.wikipedia.org/w/index.php?title=File:US_Navy_030319-N-4142G-020_Ordnance_handlers_assemble_Joint_Direct_Attack_Munition_(JDAM)_bombs_in_the_forward_mess_decks.jpg *License*: unknown *Contributors*: Avron, 1 anonymous edits

File:gps ca gold.svg *Source*: http://en.wikipedia.org/w/index.php?title=File:Gps_ca_gold.svg *License*: unknown *Contributors*: User:Tsaitgaist

Image:Comparison_satellite_navigation_orbits.svg *Source*: http://en.wikipedia.org/w/index.php?title=File:Comparison_satellite_navigation_orbits.svg *License*: unknown *Contributors*: User:Cmglee

File:Wooden hourglass 3.jpg *Source*: http://en.wikipedia.org/w/index.php?title=File:Wooden_hourglass_3.jpg *License*: unknown *Contributors*: User:S Sepp

File:Sundial Taganrog.jpg *Source*: http://en.wikipedia.org/w/index.php?title=File:Sundial_Taganrog.jpg *License*: unknown *Contributors*: EvaK, Fagairolles 34, ISasha, Kaganer, Till.niermann

File:Swatch Irony angle below.jpg *Source*: http://en.wikipedia.org/w/index.php?title=File:Swatch_Irony_angle_below.jpg *License*: unknown *Contributors*: User:Booksworm

File:ChipScaleClock2 HR.jpg *Source*: http://en.wikipedia.org/w/index.php?title=File:ChipScaleClock2_HR.jpg *License*: unknown *Contributors*: User:Lymantria

File:Signetics NE555N.JPG *Source*: http://en.wikipedia.org/w/index.php?title=File:Signetics_NE555N.JPG *License*: unknown *Contributors*: Gvf, Shooke, Stefan506, 1 anonymous edits

File:HinduMeasurements.svg *Source*: http://en.wikipedia.org/w/index.php?title=File:HinduMeasurements.svg *License*: unknown *Contributors*: BD2412, Roland zh, 1 anonymous edits

File:World line.svg *Source*: http://en.wikipedia.org/w/index.php?title=File:World_line.svg *License*: unknown *Contributors*: SVG version: K. Aainsqatsi at en.wikipedia Original PNG version: Stib at en.wikipedia

File:Relativity of Simultaneity.svg *Source*: http://en.wikipedia.org/w/index.php?title=File:Relativity_of_Simultaneity.svg *License*: unknown *Contributors*: User:Acdx

File:Lorentz transform of world line.gif *Source*: http://en.wikipedia.org/w/index.php?title=File:Lorentz_transform_of_world_line.gif *License*: unknown *Contributors*: Cyp, Daniele Pugliesi, Darapti, Jarekt, Julia W, Melchoir, Pieter Kuiper, Rovnet, Schekinov Alexey Victorovich, TommyBee, 4 anonymous edits

File:CMB Timeline300 no WMAP.jpg *Source*: http://en.wikipedia.org/w/index.php?title=File:CMB_Timeline300_no_WMAP.jpg *License*: unknown *Contributors*: GDK, Kaldari, 1 anonymous edits

File:Le Temps.JPG *Source*: http://en.wikipedia.org/w/index.php?title=File:Le_Temps.JPG *License*: unknown *Contributors*: User:Ben2

Image:Factory.svg *Source*: http://en.wikipedia.org/w/index.php?title=File:Factory.svg *License*: unknown *Contributors*: User:Howcheng

File:SCADA schematic overview-s.svg *Source*: http://en.wikipedia.org/w/index.php?title=File:SCADA_schematic_overview-s.svg *License*: unknown *Contributors*: Bender235, Bulwersator, JRaber, Xbello

Image:Scada std anim.gif *Source*: http://en.wikipedia.org/w/index.php?title=File:Scada_std_anim.gif *License*: unknown *Contributors*: Ecava

File:SCADA C4ISR Facilities.pdf *Source*: http://en.wikipedia.org/w/index.php?title=File:SCADA_C4ISR_Facilities.pdf *License*: unknown *Contributors*: USACE

Image:Siemens Simatic S7-416-3.jpg *Source*: http://en.wikipedia.org/w/index.php?title=File:Siemens_Simatic_S7-416-3.jpg *License*: unknown *Contributors*: User:Mixabest

Image:PLC Control Panel.png *Source*: http://en.wikipedia.org/w/index.php?title=File:PLC_Control_Panel.png *License*: unknown *Contributors*: Original uploader was Dailynetworks at en.wikipedia

File:BMA Automation Allen Bradley PLC 3.JPG *Source*: http://en.wikipedia.org/w/index.php?title=File:BMA_Automation_Allen_Bradley_PLC_3.JPG *License*: unknown *Contributors*: User:Elmschrat

File:Map of North America, blackout 2003.svg *Source*: http://en.wikipedia.org/w/index.php?title=File:Map_of_North_America,_blackout_2003.svg *License*: unknown *Contributors*: User:Lokal_Profil

Image:2003 North American Blackout Before.jpg *Source*: http://en.wikipedia.org/w/index.php?title=File:2003_North_American_Blackout_Before.jpg *License*: unknown *Contributors*: Original uploader was Jmorgan at en.wikipedia

Image:2003 North American Blackout After.jpg *Source*: http://en.wikipedia.org/w/index.php?title=File:2003_North_American_Blackout_After.jpg *License*: unknown *Contributors*: Original uploader was Jmorgan at en.wikipedia

File:Streetcar Toronto 2003 Blackout.jpg *Source*: http://en.wikipedia.org/w/index.php?title=File:Streetcar_Toronto_2003_Blackout.jpg *License*: unknown *Contributors*: Original uploader was Camerafiend at en.wikipedia

File:2003 New York City blackout.jpg *Source*: http://en.wikipedia.org/w/index.php?title=File:2003_New_York_City_blackout.jpg *License*: unknown *Contributors*: Edward, Infrogmation, OhanaUnited, Skeezix1000, Urban

File:toronto ON 2003 Blackout.jpg *Source*: http://en.wikipedia.org/w/index.php?title=File:Toronto_ON_2003_Blackout.jpg *License*: unknown *Contributors*: Original uploader was Camerafiend at en.wikipedia

File:Jasonp blackout2003 toronto BayAdelaide.jpg *Source*: http://en.wikipedia.org/w/index.php?title=File:Jasonp_blackout2003_toronto_BayAdelaide.jpg *License*: unknown *Contributors*: Original uploader was Jasonp at en.wikipedia

File:Jasonp blackout2003 toronto UnionStnInt.jpg *Source*: http://en.wikipedia.org/w/index.php?title=File:Jasonp_blackout2003_toronto_UnionStnInt.jpg *License*: unknown *Contributors*: Original uploader was Jasonp at en.wikipedia

File:UnitedStatesPowerGrid.jpg *Source*: http://en.wikipedia.org/w/index.php?title=File:UnitedStatesPowerGrid.jpg *License*: unknown *Contributors*: Rolypolyman (talk) Original uploader was Rolypolyman at en.wikipedia

File:NERC-map-en.svg *Source*: http://en.wikipedia.org/w/index.php?title=File:NERC-map-en.svg *License*: unknown *Contributors*: User:Bouchecl

Image:Datalogger.jpg *Source*: http://en.wikipedia.org/w/index.php?title=File:Datalogger.jpg *License*: unknown *Contributors*: Avisaro AG

File:Mini Datalogger MSR165.jpg *Source*: http://en.wikipedia.org/w/index.php?title=File:Mini_Datalogger_MSR165.jpg *License*: unknown *Contributors*: User:MSRDatenlogger

File:Data logger application for weather station.jpg *Source*: http://en.wikipedia.org/w/index.php?title=File:Data_logger_application_for_weather_station.jpg *License*: unknown *Contributors*: User:Anakwisnu

MIX
Papier aus verantwortungsvollen Quellen
Paper from responsible sources
FSC® C105338

Printed by Books on Demand GmbH, Norderstedt / Germany